Forces, Growth and Form in Soft Condensed Matter: At the Interface between Physics and Biology

T0180621

NATO Science Series

A Series presenting the results of scientific meetings supported under the NATO Science Programme.

The Series is published by IOS Press, Amsterdam, and Kluwer Academic Publishers in conjunction with the NATO Scientific Affairs Division

Sub-Series

I. Life and Behavioural Sciences	IOS Press
II. Mathematics, Physics and Chemistry	Kluwer Academic Publishers
III. Computer and Systems Science	IOS Press
IV. Earth and Environmental Sciences	Kluwer Academic Publishers
V. Science and Technology Policy	IOS Press

The NATO Science Series continues the series of books published formerly as the NATO ASI Series.

The NATO Science Programme offers support for collaboration in civil science between scientists of countries of the Euro-Atlantic Partnership Council. The types of scientific meeting generally supported are "Advanced Study Institutes" and "Advanced Research Workshops", although other types of meeting are supported from time to time. The NATO Science Series collects together the results of these meetings. The meetings are co-organized bij scientists from NATO countries and scientists from NATO's Partner countries – countries of the CIS and Central and Eastern Europe.

Advanced Study Institutes are high-level tutorial courses offering in-depth study of latest advances in a field.
Advanced Research Workshops are expert meetings aimed at critical assessment of a field, and identification of directions for future action.

As a consequence of the restructuring of the NATO Science Programme in 1999, the NATO Science Series has been re-organised and there are currently Five Sub-series as noted above. Please consult the following web sites for information on previous volumes published in the Series, as well as details of earlier Sub-series.

http://www.nato.int/science
http://www.wkap.nl
http://www.iospress.nl
http://www.wtv-books.de/nato-pco.htm

Series II: Mathematics, Physics and Chemistry – Vol. 160

Forces, Growth and Form in Soft Condensed Matter: At the Interface between Physics and Biology

edited by

A.T. Skjeltorp

Institute for Energy Technology,
Kjeller, Norway
and Department of Physics,
University of Oslo, Norway

and

A.V. Belushkin

Frank Laboratory of Neutron Physics,
Dubna, Russia

Kluwer Academic Publishers

Dordrecht / Boston / London

Published in cooperation with NATO Scientific Affairs Division

Proceedings of the NATO Advanced Study Institute on
Forces, Growth and Form in Soft Condensed Matter: At the Interface between Physics
and Biology
Geilo, Norway
24 March–3 April 2004

A C.I.P. Catalogue record for this book is available from the Library of Congress.

ISBN 1-4020-2339-1 (PB)
ISBN 1-4020-2338-3 (HB)
ISBN 1-4020-2340-5 (e-book)

Published by Kluwer Academic Publishers,
P.O. Box 17, 3300 AA Dordrecht, The Netherlands.

Sold and distributed in North, Central and South America
by Kluwer Academic Publishers,
101 Philip Drive, Norwell, MA 02061, U.S.A.

In all other countries, sold and distributed
by Kluwer Academic Publishers,
P.O. Box 322, 3300 AH Dordrecht, The Netherlands.

Printed on acid-free paper

CONTENTS

PREFACE

This volume comprises the proceedings of a NATO Advanced Study Institute held at Geilo, Norway, 24 March - 3 April 2003, the seventeenth ASI in a series held every two years since 1971. The objective of this ASI was to identify and discuss areas where synergism between modern physics, soft condensed matter and biology might be most fruitful. The main pedagogical approach was to have lecturers focussing on basic understanding of important aspects of the relative role of the various interactions - electrostatic, hydrophobic, steric, conformational, van der Waals etc. Soft condensed matter and the connection between physics and biology have been the themes of several earlier Geilo Schools. A return to these subjects thus allowed a fresh look and a possibility for defining new directions for research.

Examples of soft materials, which were discussed at this ASI, included colloidal dispersions, gels, biopolymers and charged polymer solutions, polyelectrolytes, protein/membrane complexes, nucleic acids and their complexes. Indeed, most forms of condensed matter are soft and these substances are composed of aggregates and macromolecules, with interactions that are too weak and complex to form crystals spontaneously. A characteristic feature is that small external forces, slight perturbations in temperature, pressure or concentration, can all be enough to induce significant structural changes. Thermal fluctuations are almost by definition strong in soft materials and entropy is a predominant determinant of structure, so that disorder, slow dynamics and plastic deformation are the rule. Hence the phrase 'soft condensed matter' has been coined.

Of sheer necessity, soft materials have interested engineers for ages. They also form the basis of entire fields of science such as biology. However, only recently have physicists taken an interest in such materials, and attempted to implement what is the essence of physics, that is to produce simple models that contain the irreducible minimum information required to explain essential features. One of the reasons physicists has not been able to apply that type of reasoning to biology has been the lack of data of the right sort. Only with the advent of single molecule spectroscopies and techniques such as optical tweezers are we beginning to get quantitative data on things such as DNA elasticity, the motion of molecular motors, electron and proton transfer rates etc. that are sufficiently reliable to test simple models. The availability of new experimental tools and simulation capability is allowing physicists to apply their methods to areas of biology such as the prediction of structure on the basis of known microscopic forces. This is an active area of research in which new insights are constantly emerging. For example, much of the non-specific self-assembly activity in our cells are determined by electrostatics but in a way that cannot be predicted from simple mean field theory. Examples include DNA packaging, DNA transport across membranes, virus self-assembly and some stages of protein folding. A key issue is the fact that the media in which all of this happens have strong spatial variations in dielectric constant -- e.g. water and lipids. The difficulty of this whole area is epitomized by the fact that, until recently, we couldn't even predict whether two DNA strands would attract or repel each other at close range.

The starting point, and the underlying theme throughout the ASI, was a thorough discussion of the relative role of the various fundamental interactions in soft and biological matter systems (electrostatic, hydrophobic, steric, conformational, van der

Waals, etc.). The next focus was on how these competing interactions influenced the form and topology of such systems, like polymers and proteins, leading to hierarchical structures in self-assembling systems and folding patterns sometimes described in terms of chirality, braids and knots. Finally, focus was also on how the competing interactions influence various bioprocesses like genetic regulation and biological evolution taking place in systems like biopolymers, macromolecules and cell membranes.

The scientific content of the school was timely and these proceedings should provide a useful definition of the current status.

The Institute brought together many lecturers, students and active researchers in the field from a wide range of countries, both NATO and NATO Partner Countries. The lectures fulfilled the aim of the Study Institute in creating a learning environment and a forum for discussion on the topics stated above. They were supplemented by a few contributed seminars and a large number of poster presentations. These seminars and posters were collected in extended abstract form and issued as an open report available at the Institute for Energy Technology, Kjeller, Norway (Report IFE/KR/E-2003/001).

Financial support was principally from the NATO Scientific Affairs Division, but also from the Institute for Energy Technology, NORDITA in Copenhagen and the nationally coordinated research team COMPLEX in Norway.

The editors are most grateful to J.O. Fossum, J. Hertz, M.H. Jensen, R. Pynn and H. Thomas who helped them plan the programme and G. Helgesen for helping with many practical details. Finally, we would like to express our deep gratitude to Else-Brit Jørgensen of the Institute for Energy Technology, for all her work and care for all the practical organization before, during, and after the school, including the preparation of these proceedings.

June 2003

Arne T. Skjeltorp   Alexander V. Belushkin

LIST OF PARTICIPANTS

Organizing Committee:

Skjeltorp, Arne T., Director
Institute for Energy Technology, POB 40, N-2027 Kjeller, Norway

Belushkin, Alexander V., Co-director
Frank Laboratory of Neutron Physics, Joint Institute for Nuclear Research, 141980 Dubna, Moscow region, Russia

Helgesen, Geir, Technical assistant
Institute for Energy Technology, POB 40, N-2027 Kjeller, Norway

Jørgensen, Else-Brit, Secretary
Institute for Energy Technology, POB 40, N-2027 Kjeller, Norway

Participants:

Aharony, Amnon
School of Physics and Astronomy, Tel Aviv University, 69978 Tel Aviv, Israel

Almásy, László
KFKI-SzFKI, POB 49, Budapest – 1525, Hungary

Alvarez Lacalle, Enrique
Departament ECM, Facultat de Fisica., Universitat de Barcelona, c/Marti i Franques 1 08028, Spain

Amarie, Dragos
Indiana University, Department of Physics, Swain Hall West 117, 727 E. 3rd Street, Bloomington, IN 47405-7105, USA

Antunes, Filipe
R. Virgílio Correia, lote 7, 3E, 3000 COIMBRA, Portugal

Avdeev, Mikhail
FLNP, JINR, 141980 Dubna, Moscow Reg., Russia

Bakk, Audun
Nordita, Blegdamsvej 17, DK-2100 København Ø, Denmark

Bergli, Joakim
Departement d'Estructura i Constituents de la Materia, Facultat de Fisica, Av.
Diagonal, 647, E-08028 Barcelona, Spain

Berry, Richard
Clarendon Laboratory, University of Oxford, Parks Road, Oxford OX1 3PU, United
Kingdom

Bock Axelsen, Jacob
Niels Bohr Institutet, Blegdamsvej 17, DK-2100 Copenhagen Ø, Denmark

Bogomolova, Eugenia V.
Dept. of Botany, Biology & Soil Sciences Faculty, St.Petersburg State University,
Universitetskaya emb. 7/9, 199034 St.Petersburg, Russia

Bolhuis, Peter
Dept of Chemical Engineering, University of Amsterdam, Nieuwe Achtergracht 166,
1018 WV Amsterdam, The Netherlands

Breivik, Jarle
Section for Immunotherapy, Department of Immunology, The Norwegian Radium
Hospital, 0310 Oslo, Norway

Bruinsma, Robijn
UCLA Department of Physics & Astronomy, Box 951547, Los Angeles, CA 90095-
1547, USA

Bu, Huaitian
Department of Chemistry, University of Oslo, P.O.Box 1033 Blindern, 0315 Oslo,
Norway

Buchanan, Mark
Department of Complex Systems, Division of Physics and Astronomy, De Boelelaan
1081, Amsterdam 1081HV, The Netherlands

Buyukmurat, Yasemin
Istanbul University, Science Faculty, Physics Department, Vezneciler, 34459,
Istanbul, Turkey

Campas, Otger
Dept. Estructura i Constituents de la Materia, Facultat de Fisica, Universitat de
Barcelona, Av. Diagonal, n° 647, E-08028, Barcelona, Spain

Cernak, Jozef
University of P.J. Safarik, Department of Biophysics, Jesenna 5, SK-04000 Kosice, Slovak Republic

Chertovich, Alexander
Chair of Physics of Polymers and Crystals, Physics Department, Moscow State University, Moscow 119992, Russia

Chikoidze, Ekaterine
Dipartimento di Fisica, Politecnico di Milano, Piazza Leonardo Da Vinci 32, 20132, Milano, Italy

Christophorov, Leonid
Bogoliubov Institute for Theoretical Physics, NAS Ukraine, 14 B Metrologichna Str, . Kiev 03143, Ukraine

da Silva, Geraldo José
Instituto de Fisica, Universidade de Brasilia, CP 04525 70919-970 Brasilia DF, Brazil

de Miranda Fonseca, Davi
Department of Physics, NTNU, N-7491, Trondheim, Norway

Dewimille, Philippe
Institut Curie, Section Recherche, 11, rue Pierre et Marie Curie, 75231 Paris Cedex 05, France

Dommersnes, Paul Gunnar
Physicochimie Curie, Institut Curie - Section Recherche, 11 rue Pierre et Marie Curie, F-75231 PARIS, France

Elgsaeter, Arnljot
Department of Physics, NTNU, Høgskoleringen 5, N-7491 Trondheim, Norway

Faisca, Patricia Ferreira Neves
Centro de Fisica Teorica e Computacional, Av Prof Gama Pinto 2, 1649-003 Lisbo Codex, Portugal

Fedorov, Maxim Valerievitch
Institute of Theoretical and Experimental Biophysics of Russian Academy of Sciences, Laboratory of Physical Biochemistry, Pushchino, Institutskaya str., 3, Moscow region, 142290 Russia

Finjord, Jan
Høgskolen i Stavanger, Teknat Avd., Postboks 8002, N-4068 Stavanger, Norway

Flekkøy, Eirik G.
University of Oslo, Dept. of Physics, P.O. Box 1048 Blindern, 0316 OSLO, Norway

Fossum, Jon Otto
NTNU Department of Physics, Høgskoleringen 5, N-7034 Trondheim, Norway

Futsæther, Cecilia
Dept. of Agricultural Engineering, Agricultural University of Norway, PB 5065,
N-1432 Ås, Norway

Galperin, Dmitry
Dubna, Moskow region, Tverskaya str. 15-47, Russia 141980

Giaever, Ivar
Institute of Science, Rensselaer Polytechnic Institute, Troy, NY 12180, USA

Grotkopp, Ingo
Christian-Albrechts-Universität zu Kiel, IEAP z.Hd. Herrn Grotkopp, Leibnizstrasse
19, D-24098 Kiel, Germany

Grunina, Natalia
Research Institute of Physics, St.Petersburg State University, Ulianovskaya Str., 1,
Petrodvorets 198504 St.Petersburg, Russia

Hansen, Alex
Institutt for fysikk, NTNU Gløshaugen, N-7491 Trondheim, Norway

Hauglin, Harald
Grenseveien 29, 0575 Oslo, Norway

Hawkins, Rhoda
Polymer IRC, Dept of Physics, University of Leeds, Leeds LS2 9JT, United Kingdom

Hritz, Jozef
Department of Biophysics, PF-UPJS, Jesenna 5, 04154 Kosice, Slovak Republic

Høgh Jensen, Mogens
Niels Bohr Institute, Blegdamsvej 27, DK-2100 KØBENHAVN Ø, Denmark

Israelachvili, Jacob
Department of Chemical & Nuclear Engineering, and Material Department, University
of California, Santa Barbara, California 93106, USA

Joanny, Jean-François
  Physicochimie Curie, Institut Curie Section Recherche, 26 rue d'Ulm, 75248 Paris
  Cedex 05, France

Kaya, Huseyin
  Department of Biochemistry, University of Toronto, Medical Sciences Building, 1
  King's College Circl, Toronto, Ontario M5S 1A8, Canada

Kharlamov, Alexey
  3 Krjijanovskiy str., 03142 Kiev, Ukraine

Khokhlov, Alexei R.
  Moscow State University, Physics Department, Leninskie Gory, RU-117234 Moscow,
  Russia

Kjøniksen, Anna-Lena
  Department of Chemistry, University of Oslo, P.O. Box 1033, Blindern, N-0315 Oslo,
  Norway

Knudsen, Kenneth D.
  Institute for Energy Technology, POB 40, NO-2027 Kjeller, Norway

Kristiansen, Kai de Lange
  Department of Physics, University of Oslo, P.O. Box 1048 Blindern, 0316 OSLO,
  Norway

Kölln, Klaas
  Christian-Albrechts-Universität zu Kiel, IEAP, Leibnizstrasse 19, D-24098 Kiel,
  Germany

Lomholt, Michael Andersen
  Fysisk Institut, SDU, Campusvej 55, 5230 Odense M, Denmark

Louis, Adriaan
  Cambridge University, Department of Chemistry, Lensfield Road, Cambridge CB2
  1HW, United Kingdom

Lund, Reidar
  Solid State Research (IFF), Forschungszentrum Juelich, 52425 Juelich, Germany

Manghi, Manoel
  Theorie/Sektion Physik, Ludwig-Maximilians-Universitaet, Theresienstr. 37, 80333
  Muenchen, Germany

Mañosas Castejon, Maria
Diagonal 647, Fonamental Physics Department Faculty of Physics, 08028 Barcelona, Spain

McCauley, Joseph L.
Tiefetfeld Str. 14, A6632 Ehrwald, Austria

Metzler, Ralf
Nordita, Blegdamsvej 17, DK-2100 Copenhagen Oe, Denmark

Mohrdieck, Camilla
Max-Planck-Institut für Metallforschung, Heisenbergstr. 3, 70569 Stuttgart, Gemany

Moses, Elisha
Depts. Molecular Cell Biology and Physics of Complex Systems, Weizmann Institute of Science, Rehovot, Israel 76100

Måløy, Knut Jørgen
Fysisk Institutt, University of Oslo, Box 1048 Blindern, 0316 Oslo, Norway

Nelson, David
Harvard Condensed Matter Theory Group, Harvard University, 17 Oxford Street, Cambridge, MA 02138, USA

Niemelä, Perttu Samuli
Laboratory of Physics, Helsinki University of Technology, P.O.Box 1100, FIN-02015 HUT, Finland

Næss, Stine Nalum
Department of physics, NTNU, N-7491 Trondheim, Norway

Oliveira, Fernando A.
International Center for Condensed Matter Physics, Universidade de Brasilia, CP 04513, 70919-970 Brasilia DF, Brazil

Onuta, Tiberiu-Dan
Indiana Universit , Department of Physics, Swain Hall West 117, 727 E. 3rd Street, Bloomington, IN 47405-7105, USA

Panina, Lyudmila
St.Petersburg State University, Ukhtomsky Institute of Physiology, Universitetskaya emb.7/9, 199034 St.Petersburg, Russia

Parmar, Kanak Pal Singh
Institutt for fysikk, NTNU, N-7491 Trondheim, Norway

Pépy, Gérard
  Laboratoire Léon Brillouin, CEA Saclay, 91191 Gif sur Yvette CEDEX, France

Pincak, Richard
  Watsonova 47, 043 53 Kosice, Slovak Republic

Plaxco Kevin
  Department of Chemistry and Biochemistry 9510, University of California, Santa
  Barbara CA 93106 – 9510, USA

Pogorelko, Guennady
  N.I. Vavilov Institute of General Genetics, Russian Academy of Science , Gubkin
  Str., 3, Moscow 119991 GSP-1, Russia

Pynn, Roger
  Materials Research Lab., UCSB, Santa Barbara, CA 93106-5130, USA

Rosa, Monica
  Rua da Praia, 589, 4495-031 Aguçadoura, Portugal

Rowat, Amy C.
  MEMPHYS - Center for Biomembrane Physics, Department of Physics, Southern
  University of Denmark, Campusvej 55, DK-5230 Odense, Denmark

Safinya, Cyrus R.
  Materials Research Laboratory , University of California, Santa Barbara CA 93106 –
  9510, USA

Sarvabhowman, Aravind
  SW129, Dept. of Physics, Indiana University, Bloomington, IN 47401, USA

Sewer, Alain
  SISSA, Via Beirut 2-4, 34014 TRIESTE, Italy

Shantsev, Daniel
  Department of Physics, University of Oslo, POB 1048 Blindern, NO-0316 Oslo,
  Norway

Sherrington, David
  University of Oxford, Theoretical Physics, 1 Keble Rd. Oxford OX1 3NP, United
  Kingdom

Skotheim, Jan
  Cambridge University, Queens' College, Cambridge CB3 9ET, United Kingdom

Slavchev, Radomir
Kostinbrod, Obedinena str. #55A, Bulgaria

Sletmoen, Marit
Dept. of Physics, NTNU, NO-7491 Trondheim, Norway

Sneppen, Kim
Nordita, Blegdamsvej 17, 2100 Copenhagen OE, Denmark

Spiegel, Katrin
SISSA, via Beirut 2-4, 34100 Trieste, Italy

Steinsvoll, Olav
Institute for Energy Technology, POB 40, NO-2027 Kjeller, Norway

Stibius Jensen, Karin
Niels Bohr Institutet, Blegdamsvej 17, DK-2100 Copenhagen Ø, Denmark

Thalmann, Fabrice
LDFC, Institut de Physique, 3, rue de l'Universite, 67084 Strasbourg, France

Thomas, Harry
Dept. of Physics, University of Basel, CH-4056 Basel, Switzerland

Toussaint, Renaud
Department of Physics, NTNU, N-7491 Trondheim, Norway

Tropin, Timur
FLNP, JINR, 141980 Dubna, Moscow reg., Russia

Uldina, Julia
Chair of Physics of Polymers and Crystals, Physics Department, Moscow State
University, Moscow 119992, Russia

Vilfan, Mojca
LMU Lehrstuhl Gaub, Amalienstr. 54, 80799 Muenchen, Germany

Williams, Claudine E.
Laboratoire des Fluides Organisés, CNRS URA 792, Physique de la Matière
Condensée, Collège de France, 11 place Marcelin Berthelot, 75005 Paris

Zeldovich, Konstantin
Physico-Chimie Curie, UMR 168 CNRS / Institut Curie, 11 rue Pierre et Marie Curie,
75005 Paris, France

Zinober, Rebecca Catherine
    Department of Physics and Astronomy, University of Leeds, Leeds, LS2, United
    Kingdom

THE PHYSICO-CHEMICAL BASIS OF SELF-ASSEMBLING STRUCTURES

JACOB ISRAELACHVILI
University of California
Department of Chemical Engineering
Santa Barbara, CA 93106
U.S.A.

ILYA LADYZHINSKI
Dept. of Food Engineering & Biotechnology
Technion - Israel Inst. of Technology
Technion City, Haifa 32000
Israel

Abstract: Atoms and molecules in solution can assemble into various types of 'primary' structures that can be hard (solid-like) or soft (fluid-like). Examples are small spherical micelles, cylinders (rod-like micelles), sheets (bilayers), and 'continuous' three-dimensional structures. The first lecture (Part I) describes the thermodynamic and physical principles involved in the self-assembly of soft multi-molecular aggregates. In Part II we shall see how various structural hierarchies of increasing complexity arise and can be understood by combining the thermodynamic principles laid down in Part 1 with the intermolecular and inter-aggregate forces involved. In Part III we explore more complex structures, structures that are not necessarily at thermodynamic equilibrium and that therefore do not self-assemble spontaneously, or structures that are continuously varying. Such structures have to be 'engineered', their assembly is 'directed', and they may require a continuous input of energy. We will consider their occurrence in nature (in living biological organisms) and in man-made (engineered) processes.

1. Thermodynamic and physical principles of self-assembling structures

1.1. BASIC THERMODYNAMIC EQUATIONS OF SELF-ASSEMBLY

The association of molecules into an aggregate is not entropically favorable and only occurs when it is energetically favored. The simplest case is the association of isotropic molecules such as dissolved hydrocarbons in water into large droplets which, above the solubility limit, grow into a separate phase. This association is due to the strong hydrophobic attraction between hydrocarbon molecules in water. If the hydrocarbon is in the liquid state there is no preferred structure to the aggregate other than a sphere.[1]

---

[1] A sphere has the lowest energy for an aggregate with a given number of molecules or total volume. This is because for molecules interacting isotropically with their neighbors the surface energy is the only energy that has to be minimized and a sphere has the lowest area for a given volume.

A.T. Skjeltorp and A.V. Belushkin (eds.), Forces, Growth and Form in Soft Condensed Matter: At the Interface between Physics and Biology, 1-28.
© 2004 *Kluwer Academic Publishers. Printed in the Netherlands.*

In general, the thermodynamics of self-assembly or self-association of N molecules into an aggregate can be described in terms of a 'chemical' reaction of the type:

$$A+A+A+ \ldots = A_N,$$

where the A molecules are the 'reactants' and where the 'product' is the aggregate A_N, which consists of N A-type molecules (Fig. 1). The weak, non-covalent 'bonds' that hold the molecules together do not change the thermodynamic equations; the only differences are that (i) many (usually identical) molecules appear on the left-hand side of the equation, and (ii) the binding energies are usually low, of order or less than kT per molecule (but often much more than kT for the aggregate).

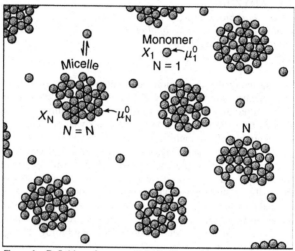

Figure 1. Definition of parameters in the thermodynamic equations for self-assembly of small, non-interacting (i.e., dilute), soft structures or aggregates in solution.

Based on the equality of the chemical potential of all molecules in the system, the basic thermodynamic equation for the distribution (or concentration) X_N of molecules in aggregates of number N is

$$X_N = N[X_1 e^{(\mu_1^o - \mu_N^o)/kT}]^N \tag{1}$$

where X_1 is the concentration of monomers (aggregates of aggregation number N=1) in solution, and where μ_1^o and μ_N^o are the interaction energies per molecule with their surroundings in the monomer and aggregated states, respectively. The total concentration C is related to X_N by

$$C = X_1 + X_2 + X_3 + \ldots + X_N + \ldots \tag{2}$$

There are many interesting consequences and implications of the above two equations for the types of aggregates formed, and it is surprising how much a purely thermodynamic equation can predict about aggregate properties, for example, why the polydispersity of 3D, 2D and 1D structures is so different [1]. These predictions are as follows:

(i)   Equation (1) shows that aggregates will form only if $\mu_N^\circ < \mu_1^\circ$ at some finite N, and

since we must have $X_1 e^{(\mu_1^\circ - \mu_N^\circ)/kT} < 1$, aggregation will occur only at concentrations C or X_1 above the so-called Critical Micelle Concentration[2] (CMC) given by

$$CMC = X_1 \; (crit) = e^{(\mu_N^\circ - \mu_1^\circ)/kT} \qquad (3)$$

in dimensionless (mole fraction) units, where we note that μ_N° is negative.

(ii)   The shape, size and polydispersity of the aggregates are intimately related and sensitive to the concentration, temperature, and solution conditions such as the ionic strength, as described below.

(iii) For very large aggregates, when N effectively becomes infinite ($N \to \infty$), 'micellization' becomes a 'phase-separation', and the CMC becomes the 'solubility limit'. But for all finite sized aggregates, even if large, the system remains a one-phase system.

(iv)   At high concentrations, inter-aggregate interactions affect μ_N° which may alter the aggregate size and shape, and ultimately leads to a phase transition to a different structure (known as 'mesophase' structures which can be interconnected in three-dimensions). As might be expected, the interaction forces between soft aggregates are intimately related to those acting between the molecules within the aggregates, both of which affect their size and shape but in different ways. This interdependence does not arise with hard colloidal particles.

For large monodisperse aggregates to form at a given aggregation number usually requires that the energy $N(\mu_N^\circ - \mu_1^\circ)$ of Eq. (1), has a minimum value at some large value of N, say N=M. This can only come about if the molecules are anisotropic, as reflected both in the shape and in their interaction with their neighboring molecules. Most amphiphilic molecules in aqueous solution possess this property, as do diblock and triblock copolymers of the type A-A-A-A-A-B-B-B-B-B-B or A-A-A-A-B-B-B-A-A-A. More complex molecules such as polysoaps, multi-headgroup or multi-chained surfactants, etc., can adopt even more complex structures.

1.2. THE ROLE OF INTERMOLECULAR FORCES AND THE MOLECULAR PACKING PARAMETER ($v/a_0 \ell_c$)

An important dimensionless parameter that often helps understand which type of structure will form is the so-called 'packing parameter', defined by ($v/a_0 \ell_c$, where v is the

---

[2] Or Critical Aggregate Concentration (CAC).

4

molecular volume, ℓ_c the maximum possible extension of the flexible chains, and a_0 is the optimum (minimum energy) head-group area at the hydrocarbon-water interface (Fig. 2). These three terms are determined either by some mechanical or steric property of the molecule (v, ℓ_c) or its interaction with the surrounding molecules (a_0) and/or the temperature (a_0, ℓ_c), as illustrated in Fig. 3.

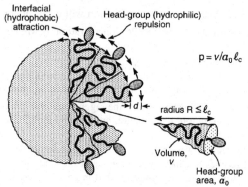

Figure 2. The surface and bulk intermolecular interactions that define the dimensionless packing parameter or average molecular 'shape factor' $v/a_0\ell_c$ of amphiphiles.

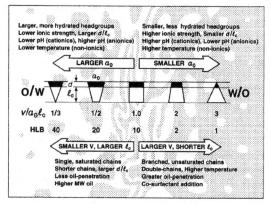

Figure 3. Factors that affect the values of v, a_0 and ℓ_c which then determine the packing parameter and the types of structures formed in water or in water-oil (W/O or O/W) mixtures. The Hydrophilic-Lyophilic-Balance (HLB) number is empirical and similar to the packing parameter but cannot be split into different, measurable contributions.

1.3. THE 'GENERIC SEQUENCE'

The packing parameter determines the types of structures formed for different molecules and solution conditions, some of which are illustrated in Fig. 4. These structures include spherical micelles (M_I), linear cylindrical micelles (H), cylindrical micelles with junctions (C), vesicles, bilayers, a variety of 'cubic' and 3D 'continuous' phases (C), lamellar structures (L), and a variety of 'inverted' structures (where the water appears to form the aggregates). These structures follow a certain order ('generic sequence') when some condition is varied, such as the head-group size, ionic strength, chain length, number of chains, temperature, water content, etc., and qualitatively similar effects occur with polymer diblocks and triblocks either in the bulk liquid or in solution.

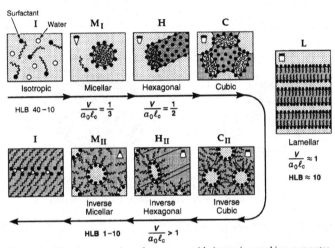

Figure 4. Generic progression of structures with increasing packing parameter (decreasing HLB number). Subscripts II refer to inverted structures. This sequence will be referred to as the 'generic sequence' of self-assembling amphiphilic structures.

The above qualitative analysis shows how the packing parameter provides a powerful qualitative tool for analyzing and predicting different structures and trends, but the full thermodynamic equations, Eqs (1) and (2), must be used whenever a complete or quantitative picture is required. This will now be done for the case of isolated structures, i.e., in dilute solutions (but above the CMC) where one can neglect short-range inter-aggregate interactions. We start by considering the five 'basic' structures, represented in the generic sequence of Fig. 4 (spheres, cylinders, lamellae, inverted cylinders and inverted spheres) and illustrated in Figure 5. The cubic, bicontinuous, tricontinuous and other 'non-standard' structures are introduced later; their structural classification generally falls somewhere in between one of the standard (basic) structures, usually between the hexagonal and lamellar or the lamellar and inverted hexagonal phases, even though some of their other properties may be very different.

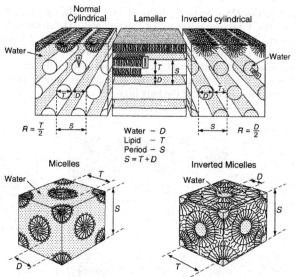

Figure 5. The five standard (basic) micellar structures: on the top are the normal cylindrical, lamellar and inverted cylindrical structures, on the bottom are the normal spherical and inverted spherical structures (micelles).

1.4. THE OPTIMAL HEADGROUP AREA (a_0)

As illustrated in Fig. 2, two opposing forces govern the self-assembly of lipids and other amphiphiles into aggregates of well defined structures [2]: a force driving the hydrocarbon chains to avoid contact with water (the hydrophobic interaction), and an opposing force between the hydrophilic headgroups driving them to make contact with water and, in so doing, effectively oppose the hydrophobic interaction. As a result of these competing forces, lipids find their headgroups interfacing with water but with the tails hidden within the aggregate's interior. Still, the two interactions are not totally satisfied – the headgroups are not totally immersed in water and the hydrocarbon chains are not totally excluded from contact with water. At the simplest level, this competition can be described by a classical model of two 'opposing forces' acting at the interface. (Actually, the two forces do not act in the same plane but are separated by a small distance d; this gives rise to an important curvature effect which is discussed later.) The hydrophobic force attempts to decrease the contact area between the tails and water which effectively acts to reduce the headgroup area a per molecule. Its contribution to the interfacial free energy per lipid can be written as γa, where γ is the effective interfacial tension of the liquid hydrocarbon-water interface (γ=30-50 mJ m^{-2}). The second force, originating from the steric and entropic repulsion (and electrostatic repulsion in the case of charged lipids) between the hydrophilic (hydrated) headgroups, tends to increase the headgroup area. In a first approximation, its contribution to the interfacial free energy can be described as k/a where k is a constant.[3]

---

[3] This is equivalent to treating the lateral headgroup interactions as in a two-dimensional van der Waals Equation of State.

The total interfacial free energy per amphiphilic (surfactant or lipid) molecule due to its intermolecular interactions with other lipids and water is therefore $F=\gamma a+k/a$, which reaches a minimum value of $F_{min}=2\gamma a_0$ at a headgroup area of $a_0 = \sqrt{k/\gamma}$, which is referred to as the *optimal headgroup area*, a_0.

In terms of the optimal area, the above expression for the free energy can be expressed as

$$F(a) = 2\gamma a_0 + \frac{\gamma}{a_0}(a - a_0)^2 \tag{4}$$

which is a parabolic function of the area about the optimal area where the energy is a minimum and equal to $2\gamma a_0$. It turns out that deviation of the headgroup area a from the optimal area a_0 is energetically rather costly. Because of that, a is usually close to a_0, regardless of the structure the lipid is in (micelle, bilayer, etc.). Equation (4) constitutes an important equation for predicting self-assembling amphiphilic structures in the absence of inter-aggregate interactions.

1.5. THE VOLUME V AND CRITICAL (FLUID HYDROCARBON CHAIN) LENGTH ℓ_c OF LIPID MOLECULES

In addition to the optimal area, each lipid is also characterized by at least two other dimensions, their volume and chain length. First, considering lipids in an aggregate as an incompressible fluid we assign to each lipid type a certain (fluid) volume v. In addition, the lipid chains can be characterized by their critical chain length, ℓ_c. The critical length sets a limit on how far the chains can extend from the headgroup, smaller extensions are allowed but further extensions are not, these being prevented by a sharp entropic rise in the aggregate energy. The length ℓ_c is a somewhat semi-empirical parameter, since it represents a vague cutoff distance beyond which chains can no longer be considered to be fluid. However, as may be expected, it is of the same order as, though somewhat less than, the fully extended molecular length of the lipid chain ℓ_{max}. For example, for a saturated hydrocarbon chain with n carbon atoms [2]:

$$\ell_c \le \ell_{max} = (0.154 + 0.126n) \text{ nm}, \tag{5}$$

and
$$v = (27.4 + 26.9n) \ 10^{-3} \text{ nm}^3. \tag{6}$$

As illustrated in Fig. 4, once the optimal area a_0, critical length ℓ_c, and lipid chain volume v are specified for a given lipid – all these being measurable or estimable – one may specify which aggregate structures will be formed in the absence of inter-aggregate forces. These structures may be conveniently specified by the dimensionless packing parameter, p

$$p = v/a_0 \ell_c. \tag{7}$$

For example, from the simple geometry of a spherical aggregate of radius R having N molecules, we have $N=4\pi R^2/a_0=4\pi R^3/3v$ and $R=3v/a_0$. Therefore, the lipids will be able to pack in the aggregate with their headgroup areas equal to a_0 and with the aggregate radius R not exceeding ℓ_c, only if

$$p < 1/3 \text{ for packing into spheres (spherical micelles)} \qquad (8)$$

In a similar fashion one obtains the following conditions for lipids to pack into the following differently shaped structures:

$$p < \tfrac{1}{2} \text{ for cylindrical structures (rod micelles)} \qquad (9)$$
$$p < 1 \quad \text{for lamellar structures (planar bilayers)} \qquad (10)$$
$$p < 2 \quad \text{for inverted cylinders (hexagonal-II phases)} \qquad (11)$$
$$p < 3 \quad \text{for inverted spheres} \qquad (12)$$

On its own, the packing parameter is not a definitive determinant of structure (it tells us more about what structures are excluded). Thus, if p<1/3, for example, all of the above structures are energetically possible (i.e., having a headgroup area of a_0). Under such conditions, the translational entropy of the aggregates now decides which of the energetically similar structures will form. The entropy will always favor the formation of the structure with the lowest aggregation number N, namely, spherical micelles in the above example. When entropic considerations are also taken into account, the above criteria may be expressed more narrowly as follows:

$$p < 1/3 \qquad \text{for spheres} \qquad (13)$$
$$1/3 < p < \tfrac{1}{2} \text{ for cylinders} \qquad (14)$$
$$1/2 < p < 1 \quad \text{for lamellae} \qquad (15)$$
$$1 < p < 3 \qquad \text{for inverted structures} \qquad (16)$$

where we recall that these aggregates must be sufficiently far apart from each other so as not to significantly interact with each other.

1.6. INCLUSION OF THE FORCES BETWEEN AGGREGATES

Inter-aggregate forces can be attractive or repulsive and can cause phase separation at higher concentrations where different structures separate out from each other. Ultimately, for any lipid-water system, at sufficiently high concentrations, the structures formed are determined by the repulsive short-range forces between the aggregates whose structures generally follow the following generic sequence with decreasing water content:

spheres (micelles) → cylinders (hex)→ lamellae (bilayers) → inverted cylinders → inverted spheres

with additional structures appearing in between some of these phases, or even replacing them. A similar sequence is found for the structures formed by polymer amphiphilies. This experimentally found sequence of 'mesophase' structures and transitions is due to the inter-aggregate forces rather than the intra-aggregate forces (which determine the primary structures formed in dilute solutions or at the CMC). Some of the most common mesophase structures are shown in Fig. 6.

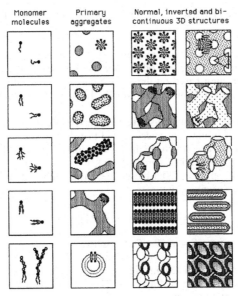

Figure 6.

In this section we consider a role of the stabilizing short-range *repulsive* interactions between aggregates. Based on experiments and theory [3, 4], the free energy per lipid due to the short-range repulsion between two aggregate surfaces separated by a water gap of thickness D (cf. Fig. 5) may be expressed simply as

$$E = E_0 \exp(-D/D_0) \tag{17}$$

where E_0 and D_0 are characteristic energy and distance parameters. Such exponential repulsions have been measured on a variety of lipid bilayer systems and often interpreted in terms of the 'hydration' repulsion between the hydrophilic (hydrated) headgroups due to structured water layers at each surface. In many cases, however, it could be due to the entropic confinement of the flexible headgroups between the surfaces [3]. Whatever its origin, we may accept the above expression as a model potential. Only at very small separations, below a few Ångstroms, does the repulsion increase more sharply than predicted by Eq. (17) – a fact that has important consequences for the preferred structures in the limit of very low water content. Adding Eq. (17) to Eq. (4) gives

$$F(a, D) = 2\gamma a_0 + \frac{\gamma}{a_0}(a - a_0)^2 + E_0 e^{-D/D_0} \tag{18}$$

which is the simplest model accounting for both the intra- and inter-aggregate interactions. Additional factors that quantitatively account the possible curvature-dependence of the interactions, the existence of a long-range *attractive* van der Waals force, and the role of entropy, will be included later.

1.7. GEOMETRIC DESCRIPTIONS OF THE FIVE BASIC STRUCTURES SHOWN IN FIG. 5

Figure 5 shows the five basic structures that we have analyzed in the first instance. The cubic, bicontinuous, tricontinuous and other 'non-standard' structures are introduced later; their structural classification based on the surface curvature generally falls somewhere in between one of the standard structures, usually between the cylindrical and inverted cylindrical phases.

It is assumed that the lamellae are stacked bilayers, that the cylindrical structures form into hexagonal arrays of infinite rod micelles, and that the spheres are arranged in a cubic close-packed lattice. For each structure, the water gap D, the lipid thickness T, and the period or spacing S are depicted in Figure 5. Note that T is the diameter of the aggregate in the case of the 'normal' structures, but the thinnest hydrocarbon thickness in the inverted structures. The spacing S is given by

$$S = T + D \text{ for cylinders and lamellae,} \tag{19}$$

whereas,
$$S = \sqrt{2}(T + D) \text{ for spheres.} \tag{20}$$

For each of the structures depicted in Fig. 5, simple geometry gives relations for the headgroup area a and lipid volume fraction f. These relations are given in Table I. We define the volume fraction of the lipids as

$$\text{Volume fraction of lipids:} \quad f = v_L N_L / (v_L N_L + v_w N_w) \tag{21}$$

where v_L and v_w, and N_L and N_w are the molecular (molar) volumes and the numbers of molecules (moles) of lipids and water, respectively, in the mixture.

TABLE I. Geometric relations for the lipid headgroup area (a) and lipid fraction (f) in terms of the lipid volume (v_l or v_L), lipid thickness (T) and water gap thickness (D).

Structure	Area a	Lipid fraction f
Lamellar	$\dfrac{v_l}{T/2}$	$\dfrac{T}{T+D}$
Cylindrical direct	$\dfrac{2v_l}{T/2}$	$\dfrac{\pi T^2}{2\sqrt{3}(T+D)}$
Cylindrical inverted	$\dfrac{2v_l}{D/2}(1/f-1)$	$1 - \dfrac{\pi D^2}{2\sqrt{3}(T+D)}$
Spherical direct	$\dfrac{3v_l}{T/2}$	$\sqrt{2}\dfrac{(4/3)\pi(T/2)^3}{(T+D)^3}$
Spherical inverted	$\dfrac{3v_l}{D/2}(1/f-1)$	$1 - \sqrt{2}\dfrac{(4/3)\pi(T/2)^3}{(T+D)^3}$

1.8. CALCULATION OF THE MINIMUM ENERGY AND AGGREGATE DIMENSIONS AT DIFFERENT LIPID FRACTIONS

Minimizing the energy Eq. (18) subject to the geometric constraints given by the above equations and those in Table I gives F, a, D, T and S as functions of the lipid fraction f for each structure. These have been computed for the different structures and are shown in Figs 7-10. In the computations, the following values of the constant parameters were used (these values were chosen as being typical for non-ionic surfactant systems [1, 2]):

Table II – Values used in calculations of lipid-water phase properties

Interfacial energy (tension) of the hydrocarbon-water interface:	$\gamma = 50$ mJ/m^2
Optimal headgroup area	$a_0 = 50\times10^{-19}$ m^2
Lipid volume (using Eq. (6) – corresponding to a lipid with two C_{18} chains):	$v_l = 10^{-27}$ m^3
Critical chain length (using Eq. (5) with n=18):	$\ell_c = 2.4$ nm (24 Å)
Critical packing parameter:	$v_l/a_0\ell_c = 0.83$
Repulsive force coefficient:	$E_0 = 5\times10^{-20}$ J (~10 kT)
Repulsive force characteristic decay length:	$D_0 = 2\times10^{-10}$ m (2 Å)
Temperature:	293 K (25°C)

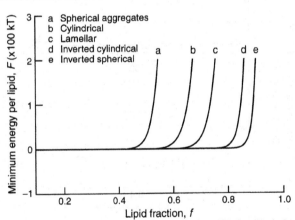

Figure 7. Computed values of the minimized energy F, Eq. (18), for spherical, cylindrical, lamellar, inverted cylindrical and inverted spherical structures, plotted as functions of the lipid volume fraction f. Zero energy corresponds to the minimum constant term $E_{min} = 2\gamma a_0$ in Eq. (4).

As can be seen in Fig. 7, increasing f beyond some value makes the free energy curves climb upwards one after the other in the order of the experimentally observed 'generic sequence'. This general result does not depend on the form of the repulsive potential used but rather to the fact that, *at any given f*, the inter-aggregate distances for different structures fall naturally in the order of the generic sequence, namely, the surfaces of spheres are closer together than the surfaces of cylinders which, in turn, are closer than the surfaces of lamellae, and so on. Thus, to minimize the unfavorable inter-aggregate interaction energy, *any* finite-ranged repulsive force will drive spheres into cylinders above some lipid concentration, then into lamellae, then into inverted structures.

1.9. EFFECT OF TRANSLATIONAL ENTROPY

Figure 7 does not include the additional effects of the translational entropy, or solvent-mixing entropy, of the differently shaped aggregates. This entropy favors the smaller aggregates. For the sake of generality, let us first assume that the packing properties of the lipids allow for all aggregates to form with the lowest energy, Eq. (4). In other words, $p \approx 1/3$. Then only the entropy will distinguish between different aggregates. Its contribution to the total free energy per lipid molecule in a dilute system of aggregates, each consisting of N molecules, can be estimated from the simple formula

$$F_{entropy} = \frac{kT}{N} \log(N/f) \tag{22}$$

Equation (22) is plotted as a function of f in Fig. 8 for N = 50, 100, 500, 1000 and 5000. Lower values of N would correspond to small, spherical aggregates (micelles), higher values would apply to cylindrical and lamellar structures.

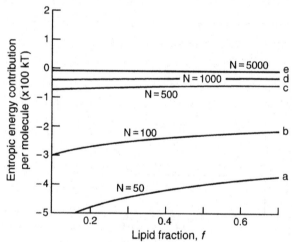

Figure 8. Contribution to the free energy per lipid due to the translational entropy (or the entropy of mixing) of aggregates as calculated from Eq. (22). The numbers refer to the numbers of lipids in the aggregates.

Adding up the free energies of Figs 7 and 8 gives Fig. 9 for the 'total' free energy per molecule within each mesophase structure, assuming an increasing number N as we go from spherical micelles to the inverted phases.

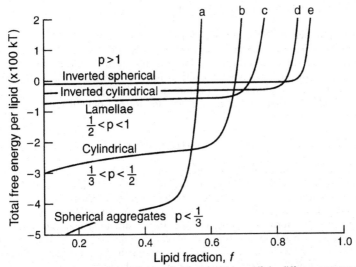

Figure 9. 'Total' free energy per lipid molecule for which p=1/3 in different aggregates, obtained by adding together the free energies of Figs 7 and 8. It is assumed that the aggregation number increases as we go from spheres to cylinders to lamellae, etc. In these computations, we have assumed N=50 for spheres, N=100 for cylinders, etc., as indicated in Fig. 8.

Figure 9 thus includes the inter-aggregate interactions and the entropy of mixing of aggregates, but assumes no packing restrictions (intra-aggregate forces) within any of the aggregate, in effect, that p≈1/3. Figure 9 shows that increasing the lipid fraction f switches the minimal free energy from one structure to another exactly in the order of the generic sequence. Therefore, Fig. 9 provides the simplest explanation of the experimentally observed trends.

Figure 9 also shows that for very large aggregation numbers, as occurs with lamellae, the effects of translational entropy become unimportant compared to the effects of the inter interaction forces. However, additional entropy terms do exist that could be important in distinguishing different aggregates, for example, those arising from undulations in bilayers, bending modes in cylinders, etc.[4]

1.10. EFFECTS OF PACKING CONSTRAINTS

We now consider the effects of having $p>1/3$. If, for example, $p=1/2$, the curve for spheres ($p=1/3$) in Fig. 9 will be shifted upwards to higher (less favorable) energies, but all the other curves will remain unchanged. This will therefore eliminate spheres from the sequence, which will now start with cylindrical micelles already at the CMC, and then proceed along the 'generic sequence' as before. Likewise, if $p>1/2$, the first structures to form at the CMC will be lamellar (e.g., vesicles), and for $p>1$ only inverted structures will occur.

1.11. AGGREGATE DIMENSIONS: SIZE, THICKNESS, ETC.

We now turn our attention to the surface area per lipid a, the water gap thickness D, the lipid thickness T, and the overall repeat spacing S, all of which were computed during the minimization of Eq. (18) for the free energy. The computed values for the five standard structures are plotted in Fig. 10 (a to e) again using the constants listed in Table II. The figures provide rich information about changes in the structural parameters within each mesophase that, in general, follow the trends observed experimentally [6-10]. One can see, for example, how increasing the lipid fraction *increases* the lipid thickness in the normal and lamellar mesophases, but causes a *decrease* in inverted phases.

---

[4] The above treatment of the entropy factor is highly simplified. To make the formula as simple as Eq. (22), we assume that the volumes of water and lipids molecules are the same. However, the main simplification results from the fact that the number of molecules in lamellar and inverted aggregates, while large, is not well defined. The effective aggregation numbers and the entropy of these structures are rather controlled by defects and fluctuations in the aggregates whose mechanisms are still not fully understood. Some relevant ideas about these mechanisms have been suggested in the theoretical works of Gelbart et al. [5]. It is also worth noting that some of the predicted trends of the above model, which is based on the repulsive inter-aggregate forces, can also be accounted for by considering the relative configurational entropies of spheres, cylinders and lamellae. This entropic contribution is different from the translational entropy of the aggregates, as defined by Eq. (22), and is small compared to the energetic contribution.

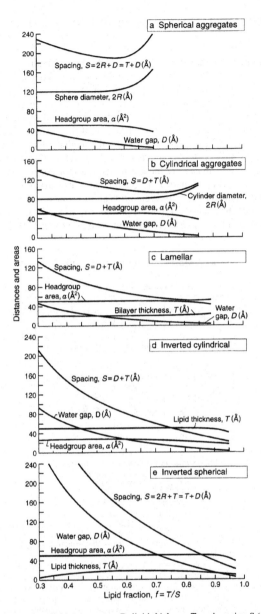

Figure 10. Area per lipid a, water gap D, lipid thickness T, and spacing S (as defined in Fig. 5) for (a) spheres, (b) cylinders, (c) lamellae, (d) inverted cylinders, and (e) inverted spheres. For each aggregate, the computed dimensions correspond to the minimum free energy state. The equations used are listed in Table I and include Eqs (18)–(21).

16

1.12. 'NON-STANDARD' PHASES: CUBIC, BICONTINUOUS, TRICONTINUOUS, ETC.

To obtain a complete picture, and to explain why non-standard and out-of-sequence structures often occur, we also have to consider (i) the curvature contributions to the energy, (ii) the contributions from the attractive (e.g., van der Waals) forces between aggregates, and of course (iii) other types of structures (cf. Fig. 6) and their energetics.

According to Luzzatti there are six different cubic phases in lipid-water systems whose structures are firmly established [8]. Of these, three are of the normal micellar and one is of the inverted micellar type. One of the three normal micellar phases corresponds to our model of close-packed (body centered) spheres, which we referred to as 'normal spherical'. The other 'established' cubic phases are the 3-arm and 4-arm tricontinuous phases (Fig. 11). The corresponding space group symbol is **Ia 3d** for the 3-arm structure and **Pn 3m** for the 4-arm structure. Water is inside the cylindrical arms and for this reason these two phases are always referred to as of the inverted type. Regarding the tricontinuous 6-arm structure of space group **Im 3m** (Fig. 11), its existence has remained controversial [cf. ref. 8 and refs 1-3 quoted therein].

NON-STANDARD (CUBIC) STRUCTURES

Bicontinuous structures

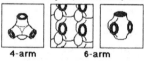

4-arm normal   inverted   6-arm normal ≡ inverted

Tricontinuous bilayer (membrane) structures

4-arm   6-arm

Tricontinuous cylindrical structures with 3 or 4-arm junctions

Normal   Inverted

Figure 11. The tricontinuous 3-arm, 4-arm and 6-arm cubic structures. In the literature, these structures are always referred to as *bicontinuous* even though they have *three* unconnected continuous zones. True bicontinuous structures also exist (cf. Fig. 6). Here, we distinguish between tricontinuous and bicontinuous structures. The geometry of the tricontinuous structures, modeled as interconnected truncated cylinders was first analyzed in ref. [9]. In Table III, we have updated and expanded the data reported in ref. [9]. While the experimental data on these structures could be more or less reliably extracted from the x-ray data (except for the 'unreliable' 6-arm case), many important questions, e.g., the real shapes of the channels (assumed cylindrical), could not be experimentally resolved [8-10].

The periods of spacing S of the 3-arm, 4-arm and 6-arm tricontinuous cubic structures are depicted in Fig 11. In the case of the 4-arm cubic, Fig. 11 shows more structure than the periodical unit cell (of volume S^3) to provide a better idea about the structure. The water gap D and the lipid thickness T_{min} (as detailed below) are depicted for the 6-arm structure. The arms (channels) of the structures are filled with *water* and are connected in a 3-dimensional network. In each structure, there are two such water networks which are not connected one to another. The space between the two networks is filled with a lipid 'bilayer' which forms the third 3-dimensional lipid network (hence tricontinuous).

For each structure, Table III gives the angles formed between the arms, the total number of water channels N_c per unit cell (each channel shares two arms), and the value of the ratio:

$$\frac{L}{S} = \frac{\text{length of the axis of the channel}}{\text{spacing}} . \tag{23}$$

To proceed, we assume that each channel is a cylinder that is truncated to meet the other two, three or five cylindrical arms. With D being the water gap diameter of each truncated cylinder, its volume and the area interfacing with the lipid headgroups take the following values:

Volume of a channel $\qquad V = \dfrac{\pi D^2}{4} L(1 - k_v \dfrac{D}{2L})$ $\qquad\qquad$ (24)

Interfacial area per channel $\qquad A = \pi DL (1 - k_A \dfrac{D}{2L})$, $\qquad\qquad$ (25)

where the constants k_v and k_A are given in Table III. Given the formula for the lipid fraction

$$f = 1 - N_c V / S^3 \tag{26}$$

the above equations yield the following relations between the area per lipid headgroup a, the lipid fraction f, the spacing S, the water gap D, and the channel length L:

Headgroup area $\qquad a = \dfrac{2v_1}{D/2} (1/f - 1) \dfrac{1 - k_A D/2L}{1 - k_v D/2L}$ $\qquad\qquad$ (27)

Lipid fraction $\qquad f = 1 - N_c \pi (1 - k_v \dfrac{D}{2L}) \dfrac{DL^2}{S^3}$. $\qquad\qquad$ (28)

Note the similarity between Eq. (27) and the equation for the headgroup area a of the inverted cylindrical structure in Table I. The minimal and maximal lipid thicknesses (note that in these structures the lipid thickness is not uniform) are obtained via figuring out the separations between the axes of the two unconnected water networks:

Lipid thickness $\qquad T_{min} = 2k_{min}S - D$ $\qquad\qquad$ (29)

$\qquad\qquad\qquad\qquad T_{max} = 2k_{max}S - D$ $\qquad\qquad$ (30)

where the coefficients k_{min} and k_{max} are also given in Table III.

Table III. Geometrical relations for the three cubic structures.

Number of arms	3	4	6
Angle between arms	120°	thetrahedric angle	90°
Number of channels per unit cell, N_c	24	4	6
Ratio $\dfrac{L}{S}$, Eq. (23)	$1/\sqrt{8}$	$\dfrac{\sqrt{3}}{2}$	1
Constant k_v in Eq. (24)	0.49	0.78	1.20
Constant k_A in Eq. (25)	0.735	1.17	1.80
Constant k_{min} in Eq. (29)	$\dfrac{\sqrt{3}}{8}$	$1/\sqrt{8}$	1/4
Constant k_{max} in Eq. (30)	$\dfrac{\sqrt{5}}{8}$	$1/\sqrt{6}$	$1/\sqrt{8}$

1.13. CALCULATION OF THE MINIMUM FREE ENERGIES AT DIFFERENT LIPID-WATER FRACTIONS

Following the same procedure as for the five 'basic' structures, a free energy minimization of Eq. (18) subject to the constraints given by Eqs (23) and (27)-(30) and those of Table III gives us the free energies F and the spatial dimensions, S, D, T_{min} and T_{max} of the three cubic structures. Using the same constants as listed in Table II, we plot in Fig. 12 the calculated free energies against the background of the free energies of the basic structures. As can be seen, the three cubic phases fall very close together and between the lamellar and inverted cylindrical structures, but much closer to the latter.

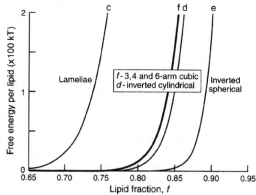

Figure 12. Computed values of the minimized free energy F (Eq. (18)) as a function of lipid fraction f, for the 3-arm, 4-arm and 6-arm tricontinuous cubic structures, and the corresponding curves for the cylindrical, lamellar and inverted cylindrical structures. (The last three curves are the same as in Fig. 7 but plotted on a different scale). Zero free energy corresponds to the constant term $E_{min} = 2\gamma a_0$ of Eq. (18).

The model of the bevel-shaped cylindrical arms yields the following important result that **with regard to the short-range inter-aggregate repulsion, the three tricontinuous cubic structures are hardly distinguishable.** Additional factors, such as curvature effects and/or long-range *attractive* inter-aggregate surface forces, must therefore distinguish between these structures. In the last section we end by analyzing these effects.

1.14. CURVATURE ENERGY CONTRIBUTIONS

As was mentioned above in the description of the model of 'opposing forces', these two forces are not applied at the same plane. The attractive hydrocarbon-water interfacial tension must be acting somewhere closer to the aggregate's interior than the headgroup repulsion which acts a distance d above the interface (cf. Figs 2 and 3). A moment created by the two forces causes a dependence of the free energy on the local curvature of the interface, which we refer to as the curvature effect. These two 'opposing forces' are not the only possible contributors to a curvature effect. One can also expect a contribution from the repulsive pressure of the lipid tails acting within the aggregate, i.e., below the interface. Thus d could be positive or negative. A molecular description of the curvature effect is not yet available; however, phenomenologically it can be modeled as the resulting moment of the same two 'opposing forces' as before, but now acting in different planes separated by a small distance d, which can be positive or negative. The curvature effect is inactive in planar lamellae due to their infinite radius of curvature, but arises in spherical, cylindrical and other complex (non-zero curvature) structures.

For curved structures of radius R (R=T/2 for normal structures, R=D/2 for inverted structures) it can be shown [1] that the contribution of the two forces to the free energy per lipid can be described by expressions similar to Eq. (4), with the same constants, but where the areas a are modified to account for the finite d value as follows:

$$F(a) = \gamma a (1 \pm d/R)^n + k/a \quad \text{for normal structures} \tag{31}$$

$$F(a) = \gamma a + k/a (1 \pm d/R)^n \quad \text{for inverted structures} \tag{32}$$

where $n=1$ for cylindrical aggregates and n=2 for spherical aggregates. The two expressions reach their minimums (optimum areas) at

$$a = a_0 (1 \pm d/R)^{-n/2} \tag{33}$$

and the free energy near the minimums are

$$F = 2\gamma a_0 (1 \pm d/R)^{n/2} + \frac{\gamma}{a_0} (1 \pm d/R)^{3n/2} [a - a_0 (1 \pm d/R)^{-n/2}]^2 \quad \text{for normal structures} \tag{34}$$

and $F = 2\gamma a_0 (1 \pm d/R)^{-n/2} + \dfrac{\gamma}{a_0} (1 \pm d/R)^{n/2} [a - a_0 (1 \pm d/R)^{-n/2}]^2$ for inverted structures (35)

where for both types of structures $a_0 = \sqrt{k/\gamma}$, as before. Equations (34) and (35) transform back to Eq. (4) in the absence of curvature effects (d=0) or in the case of planar lamellar

20

structures (R=∞). Finally, adding the repulsive inter-aggregate potential $E_o e^{-D/D_0}$ of Eq. (17) gives the full expressions for the free energies

A reasonable estimate for the separation distance d between the two forces is a few Ångstroms. In this report we restrict the calculations to the inverted cylindrical and spherical structures only, and setting d at ±0.5Å. The minimization of Eq. (35) is performed subject to the usual constraints of the structures.

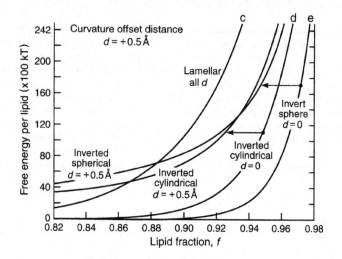

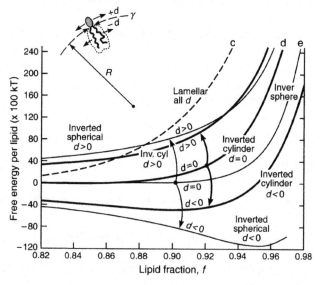

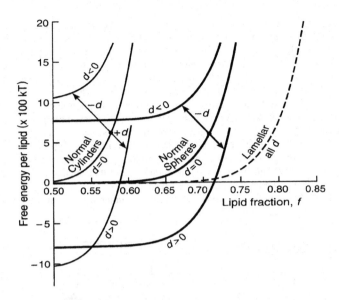

Figure 13. Computed values for the minimum free energy, given by Eqs (31)-(35), which take into account the curvature energy contribution by setting the interface width (hydrophobic-hydrophilic separation distance) d at 0, +0.5Å or -0.5Å. Note that for lamellae the curvature effect is inactive because R=∞). Zero free energy corresponds to the constant term $E_{min} = 2\gamma a_0$. It is remarkable how such a small value for d can have such a large effect.

Thus calculated, the 'curvature-modified' free energies as a function of lipid fraction are shown in Fig. 13, which show that the inverted spherical structure, which was more favorable than the cylindrical one before accounting for the curvature contribution, becomes less favorable at practically all lipid fractions once d exceeds +0.5Å! The lamellar structure, which was always less favorable than the two inverted structures, becomes more favorable at a low lipid fraction and, over a certain range, falls in between the two. (The translational entropy contributions discussed earlier are negligible at the lipid fractions shown in Fig. 13).

Figure 13 suggests a number of possible reasons for why some surfactants or lipids, depending on their headgroup length and chain fluidity, both of which affect d, may deviate from the 'generic sequence' of mesophase transitions.

2. Intermolecular, interparticle and surface forces involved in self-assembly

Most self-assembly processes occur in solution, usually an aqueous solution containing inorganic ions and organic solutes such as surfactants, lipids or polymers. The forces between the assembling molecules within each aggregate determine the primary structures they adopt, as described in Part I. But these structures also interact with each other (via the same or other forces) in the solution, and these in turn lead to 'higher-order' structures or

22

assemblies as described in the latter part of Part I. This hierarchical process is analogous to the primary, secondary, tertiary, quaternary, and higher order structures of proteins and filamentous or rod-like structures such as DNA and actin, but it involves different molecules and forces, and the resulting structures are usually softer, i.e., 'fluid-like'. The forces themselves involve van der Waals, electrostatic, steric, entropic and solvation forces, as well as hydrogen-bonding and hydrophobic interactions which are still not fully understood. In Part I, these forces were treated very simply and phenomenologically. Here we briefly consider them in more details and how they affect self-assembly processes.

Inter-aggregate forces become progressively more important as the structures are forced closer together, i.e., as the total amphiphile concentration increases or as the water content diminishes. This is particularly important in biological situations where the water content is typically ~70% and where most membranes and other surfaces are never more than ~10 nm away from another membrane or surface.

Figures 14 and 15 show the main types of forces encountered between soft amphiphilic and biological surfaces in aqueous solutions.

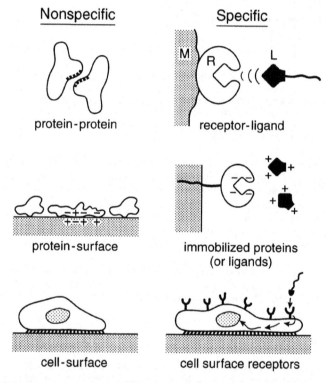

Figure 14. The long-range colloidal and short-range adhesion or steric forces between surfaces in water can be non-specific (as in the case of van der Waals, hydrophobic and electrostatic double-layer forces), or specific as in the case of short-range discrete ionic and lock-and-key type interactions.

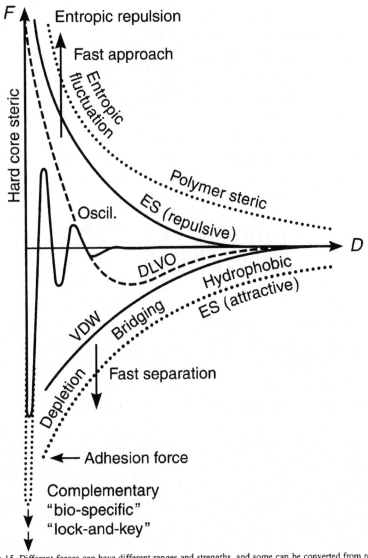

Figure 15. Different forces can have different ranges and strengths, and some can be converted from repulsion to attraction or vice versa. Figure adapted from ref. [11].

When acting together, these forces are not simply additive, but can act in very subtle ways, having different effects that are a functions of both space and time. Some of these are illustrated in Figs 16 and 17.

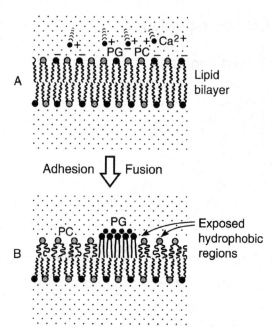

Figure 16. Example of great subtlety resulting from a combination of electrostatic and ionic interactions acting on and between two mixed lipid bilayers composed of charged and uncharged lipids. Calcium ions cause condensation of the negatively charged lipids which forces the reservoir of uncharged lipids to expand, exposing more of their hydrophobic chains to the aqueous phase, and thereby causing adhesion and fusion of these regions to another, nearby bilayer.

3. Equilibrium versus non-equilibrium structures: 'direct assembly'

Some self-assembly processes are thermodynamically driven and therefore occur spontaneously, giving rise to 'spontaneous assembly', even though the equilibrium state may be reached very slowly. Others must be engineered or directed by a judicious input of energy or work. This may be referred to as 'directed assembly'. Spontaneous assembly results in equilibrium structures, directed assembly does not. Most biological structures are non-equilibrium, either changing continuously or remaining unchanged in some steady-state configuration that requires a continuous input of energy (e.g., via ATP) and that is therefore not the 'thermodynamically' equilibrium state.

Interactions that are in a steady-state or evolving in space or time must be considered as 'processes' rather than as 'an interaction'. They have not real beginning nor end, each elemental step proceeding sequentially as the molecules move about within the system. The overall picture may be very complex (see, for example, the transport process depicted in Fig. 14, bottom right) but, like all biological processes, they contain great order and efficiency.

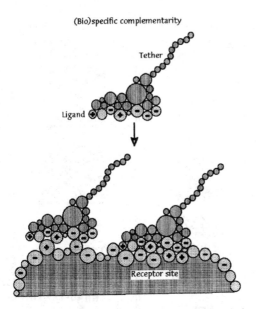

(Bio)specific complementarity

Tether

Ligand

Receptor site

Figure 17. Specific ionic binding resulting from two overall negatively charged surface regions (a negatively charged ligand near a negatively charged protein surface) that repel each other at large separations where the charges are seen as effectively smeared out, but strongly attract on contact when opposite charges come together, which occurs only in one unique spatially specific configuration.

Our attempts to mimic such systems should allow us to develop various engineered systems, as a well as 'smart', 'tunable' and 'adaptable' systems, as illustrated in figs 18-20.

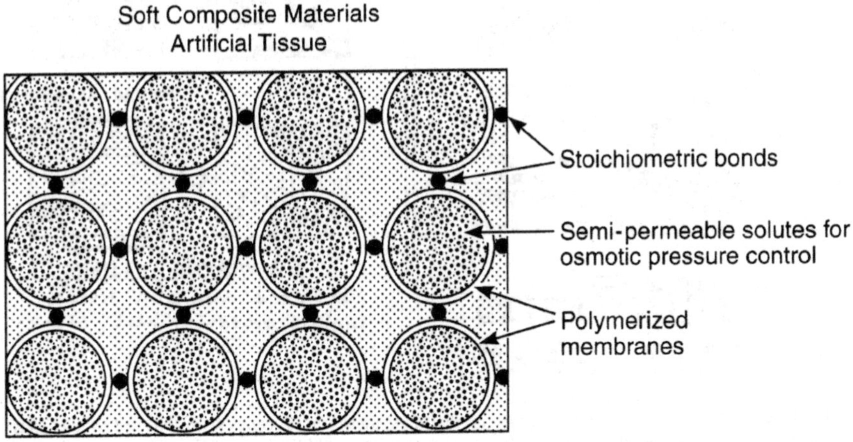

Soft Composite Materials
Artificial Tissue

Stoichiometric bonds

Semi-permeable solutes for osmotic pressure control

Polymerized membranes

Figure 18. Bioengineered soft structural composite material or artificial tissue.

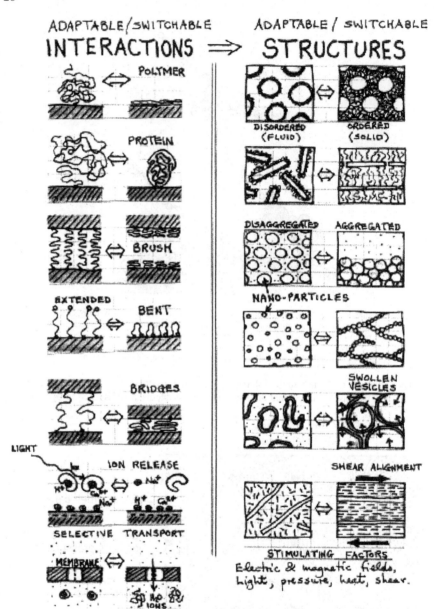

Figure 19. Different types of tunable interactions that could transform structures when activated by light, pressure, heat or an electric or magnetic field [12]. The altered structures could have very different mechanical and electro-optical properties.

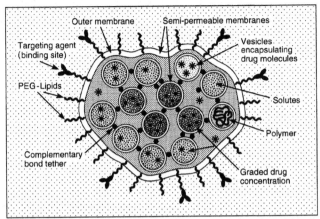

Figure 20. Drug-delivery system (vesosome) of the future where an artificially created biomimetic cell, injected into the body, will selectively target an infected tissue. Upon binding to the tissue, the 'vesosome' will become internalized where it will slowly release the drug.

Acknowledgement: This work was supported by the MRSEC Program of the National Science Foundation under Award No. DMR00-80034.

References

1. J. N. Israelachvili, D. J. Mitchell & B. W. Ninham (1976) Theory of Self-Assembly of Hydrocarbon Amphiphiles into Micelles and Bilayers, J. Chem. Soc. Faraday Trans. II, 72, 1525-1568.
2. C. Tanford (1980) The Hydrophobic Effect, Wiley, New York.
3. J. N. Israelachvili & H. Wennerström (1996) Role of Hydration and Water Structure in Biological and Colloidal Interactions, Nature 379, 219-225.
4. J. N. Israelachvili (1991) Intermolecular and Surface Forces (2nd Edition), Academic Press, London & New York.
5. W. M. Gelbart, A. Ben-Shaul (1996) The "New" Science of "Complex Fluids", J. Physical Chemistry, 100, 13169-13189.
6. J. M. Seddon (1990) Structure of the Inverted Hexagonal (HII) Phase and Non-lamellar Phase Transitions of Lipids, Biochimica et Biophysica Acta, 1031, pp 1-69.
7. R. P. Rand and N. L. Fuller (1994) Structural Dimensions and Their Changes in Reentrant Hexagonal-Lamellar Transitions of Phospholipids, Biophysical Journal 66, 2127-2138.
8. V. Luzzatti, H. Delacroix and A. Gulik (1996) The Micellar Cubic Phase of Lipid-Containing System, J Phys. II France 6, 405-418.
9. A. Gulik and V. Luzzatti (1985) Structure and Polymorphism of Bipolar Isopranyl Ether Lipids from Archaebacteria, J. Mol. Biol., 182, 131-149.
10. V. Luzzatti (1968) X-ray Diffraction Studies of Lipid-water Systems, in: Biological Membranes, Vol. 1, ed. D. Chapman (Academic Press, London) pp. 71-123.

11. D. Leckband and J. Israelachvili (2001) Intermolecular forces in biological systems, Quart. Revs. Biophys., 34, 105-267.
12. J. Lahann et al. (2003) A reversibly switching surface, Science 299, 371-374.

SUPRAMOLECULAR ASSEMBLY OF BIOLOGICAL MOLECULES

DNA-lipid complexes and gene therapy; cell cytoskeletal assembly in vitro

CYRUS R SAFINYA

Materials and Physics Departments, Biomolecular Science & Engineering Program
University of California at Santa Barbara
Santa Barbara, California 93106 USA

ABSTRACT In the first part of these lectures we describe recent work on the self-assembled structures of cationic liposome-DNA (CL-DNA) complexes by the quantitative techniques of synchrotron x-ray diffraction. CL-DNA complexes are currently used in worldwide gene therapy applications. Unraveling the relation between structure and transfection efficiency (i.e. DNA transfer and expression) is the goal of such research. Distinct structures have been discovered including, a multilamellar structure with alternating lipid bilayer and DNA monolayers, and an inverted hexagonal structure with DNA coated by lipid tubules. Confocal optical imaging and cell transfection results are beginning to unravel the relationship between these nanostructured assemblies and transfection efficiency. In the last part we briefly present a discussion of filamentous actin-cytoskeletal assembly in vitro.

1. Background and Significance (DNA-lipid complexes and gene therapy)

DNA chains dissolved in solution are known to give rise to a rich variety of condensed and liquid crystalline phases at high concentrations [1,2]. More recently there has been experimental and theoretical work on DNA chains mixed with oppositely charged cationic lipids. Early on oligo-lamellar structures had been reported in cryo-TEM studies [3]. Freeze fracture electron microscopy study had also observed isolated DNA chains coated with a lipid bilayer [4]. The dominant structure which forms spontaneously when DNA (a negative polyelectrolyte) is complexed with cationic liposomes (CLs) (closed bilayer shells of lipid molecules, Fig. 1) containing a specific type of lipid is a multilayer assembly of DNA sandwiched between bilayer membranes shown schematically in Fig. 2 [5,6]. As we describe synchrotron x-ray scattering [5,7] also shows that linear DNA confined between bilayers forms an expanding one-dimensional (1D) lattice of chains; that is, *a novel 2D smectic phase* resulting from long-range electrostatic repulsions. The CL-DNA complex thus consists of a new "hybrid" phase of matter; that is, where the DNA chains form a finite size two-dimensional (2D) smectic coupled to the three-dimensional (3D) smectic lamellar phase of lipids.

A.T. Skjeltorp and A.V. Belushkin (eds.), Forces, Growth and Form in Soft Condensed Matter: At the Interface between Physics and Biology, 29-50.
© 2004 *Kluwer Academic Publishers. Printed in the Netherlands.*

The structure and thermodynamic stability of these CL-DNA complexes has also been the subject of much recent theoretical work [8-14]. Analytical and numerical studies of DNA-DNA interactions bound between membranes show the existence of a novel long-range repulsive electrostatic interaction [10-12]. Theoretical work on CL-DNA complexes has also led to the realization of a variety of novel new phases of matter in DNA-Lipid complexes [13,14]. In particular, a novel new "sliding columnar phase", which remains to be discovered experimentally, is found where the positional coherence between DNA molecules in adjacent layers is lost without destroying orientational coherence of the chains from layer to layer. This new phase would be a remarkable new phase of matter if it exists and shares many fascinating similarities with flux lattices in superconductors. From a fundamental soft condensed matter perspective DNA and polyelectrolyte biopolymers (F-actin and microtubules) confined to the surface of membranes are models for studies of the statistical mechanics of polyelectrolytes in 2D.

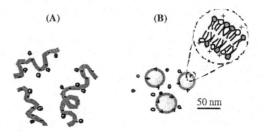

(A)   (B)

50 nm

Fig.1 (A) DNA with cationic counter-ions condensed on the backbone due to Manning condensation. (B) Cationic liposomes (spherical membranes or vesicles) containing a bilayer of a mixture of cationic and neutral lipids. (Bar for (B))

In its own right from a biophysics perspective, it is important to explore the phase behavior of DNA in two dimensions as a tractable experimental and theoretical system for understanding DNA condensation. The mechanisms of DNA condensation in vivo (i.e. packing in a small space) are poorly understood [15]. DNA condenstion and decondensation, which happens, for example, during the cell-cycle involves different types of oppositely charged polyamines, peptides and proteins (e.g. histones) where the nonspecific electrostatic interactions are clearly important. In bacteria which are the simplest cell types, it is thought that multivalent cationic polyamine molecules (spermine, spermidine) are responsible for DNA condensation in the 3D space of the cell cytoplasm. Recent experiments have led to a novel new finding where we observe that at a critical concentration of divalent cations, the forces between DNA chains in CL-DNA complexes, reverses from repulsive to attractive leading to a DNA condensed phase in 2D [16]. The importance of the observation lies in the fact that in vitro in 3D, DNA solutions containing divalent cations exhibits the usual repulsive interactions with no hint of attractive interactions. Thus, it appears that the strength of the attractive Coulomb interactions between similarly charged polyelectrolyte rods is a strong function of either the dimension in which the rods reside or the effects of confinement between lipid bilayers.

From a technological perspective once methods have been developed to produce highly oriented single crystals, DNA-membrane self-assemblies will be suitable for the development of nano-scale masks in lithography and molecular sieves with nanometer scale cylindrical pores in separations technology [17,18]. Finally, from a biomedical and biotechnological point of view, cationic liposomes (or vesicles) are empirically known to be efficient synthetic carriers of genes (i.e. sections of DNA) in synthetic gene delivery applications [19-27]. It is only recently that we have experimentally discovered the new types of liquid crystalline self assembled structures in these biomedically important membrane-polyelectrolyte complexes.

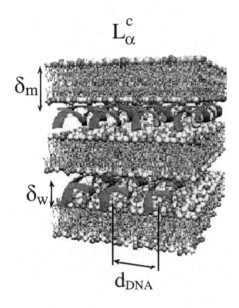

Fig. 2 The lamellar L_α^C phase with alternating lipid bilayer-DNA monolayer of cationic lipid-DNA (CL-DNA) complexes as described in the text. The interlayer spacing is $d = \delta_w + \delta_m$. Model is based on synchrotron x-ray diffraction data (From reference [5], also see reference 7)

2. Experiments on the Lamellar L_α^C Phase of Cationic Lipid –DNA Mixtures

We have carried out a combined *in-situ* optical microscopy and x-ray diffraction study of CL-DNA complexes [5,7] where the cationic liposomes consisted of mixtures of neutral (so called "helper-lipid") DOPC and cationic DOTAP (Fig. 3).   High resolution small angle x-ray diffraction has revealed that the structure is different from the hypothesized "bead-on-string" structure, originally proposed by Felgner et al. for CL-DNA complexes in their seminal paper [28], picturing the DNA strand decorated with distinctly attached cationic liposomes (refer to Fig. 1). The addition of linear λ -phage DNA (λ-DNA, 48,502 bp, contour length = 16.5 μm) to binary mixtures of cationic liposomes (mean diameter of 70 nm) induces a topological transition from liposomes into collapsed condensates in the form of optically birefringent liquid crystalline globules with size on the order of 1 μm. Fluorescent microscopy of the DNA (labeled with YoYo) and the lipid (labeled with Texas Red-DHPE) also showed that the individual globules contain both lipid and DNA [5,29]. Polarized microscopy also shows that the distinct globules are birefringent indicative of their liquid crystalline nature.

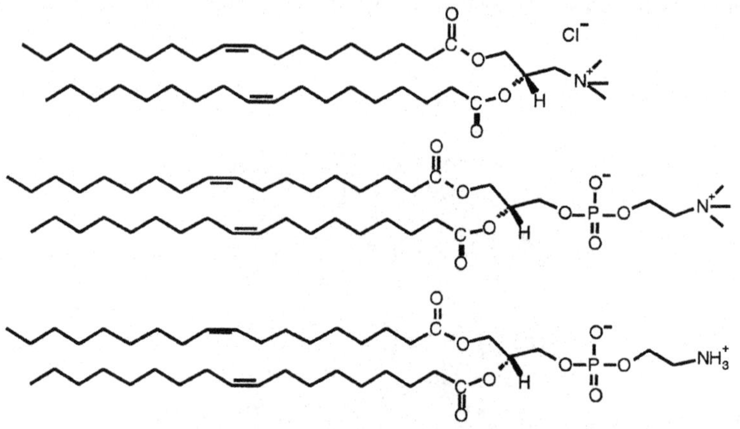

Fig. 3. TOP: The cationic lipid DOTAP (dioleoyl trimethylammonium propane). MIDDLE: The neutral helper-lipid DOPC (dioleoyl-phosphatidylcholine). BOTTOM: The neutral helper-lipid DOPE (dioleoyl-phosphadtidylethanolamine). The neutral lipid is referred to as the "helper lipid" as it tends to help cell transfection (see references 21 and 22).

The charge of the complexes was measured by their electrophoretic mobility in an external electric field [5]. For L/D > 5 the complexes are positively charged, while for L/D < 5 the complexes are negatively charged. The charge reversal is in good agreement with the stoichiometrically expected charge balance of the components DOTAP and DNA at L/D ≈ 4.4 (Wt./Wt.) where L = DOTAP + DOPC in equal weights. Thus, the positively and negatively charged globules at L/D = 50 and L/D = 2 respectively, repel each other and remain separate, while as L/D approaches 5, the nearly neutral complexes collide and tend to stick due to van der Waals attraction

The small-angle-x-ray scattering (SAXS) experiments [5,7,29] revealed a novel self assembled structure for the condensed globules consisting of mixtures of CLs and DNA. Fig. 4 is a plot of SAXS data of λ-DNA - DOPC/DOTAP complexes as a function of increasing Φ_{DOPC} (the weight fraction of DOPC in the DOPC/DOTAP cationic liposome mixtures). The data are consistent with a complete topological rearrangement of liposomes and DNA into a multilayer structure with DNA intercalated between the bilayers (Fig. 2, denoted L_α^c). To see this we first consider complexes of DNA and DOTAP at $\Phi_{DOPC} = 0$ (Fig. 4 bottom). The two sharp peaks at $q = 0.11$ and 0.22 Å^{-1} correspond to the (00L) peaks of a layered structure with an interlayer spacing d ($= \delta_m + \delta_w$) = 57 Å. The membrane thickness and water gap are denoted by δ_m and δ_w, respectively (Fig. 2).

In the absence of DNA, membranes of the cationic DOTAP exhibit strong long-range interlayer electrostatic repulsions that overwhelm the van der Waals attraction. In this case, as the volume fraction Φ_w of water is increased, the L_α phase swells and the intermembrane distance d (which is measured by SAXS) is given by the simple geometric relation $d = \delta_m/(1-\Phi_w)$. For membranes of pure DOTAP $\delta_m = 33 \pm 1$ Å [5]. Highly dilute liposomes of DOTAP (with $\Phi_w \approx 98.5$ % used in the SAXS experiments) do not exhibit Bragg diffraction in the small wave-vector range covered in Fig. 4.

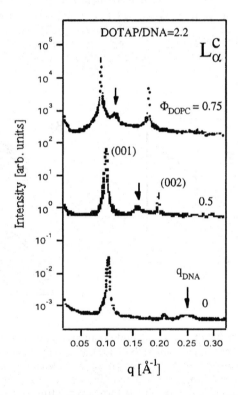

Fig. 4 SAXS scans of CL-DNA complexes at constant DOTAP / DNA = 2.2 (at the isoelectric point) with increasing DOPC / DOTAP which shows the DNA peak (arrow) moving towards smaller q, indicating that d_{DNA}

increases, as L/D (and Φ_{DOPC}) increases. (L = DOTAP+DOPC, D = DNA) (adapted from references 5 and 29).

Thus, the DNA that condenses on the cationic membranes strongly screens the electrostatic interaction between lipid bilayers and leads to condensed multilayers. The average thickness of the water gap $\delta_W = d - \delta_m = 57$ Å $- 33$ Å $= 24$ Å ± 1 Å is just sufficient to accommodate one monolayer of B-DNA (diameter ≈ 20 Å) including a hydration shell. As we now discuss the broad peak denoted $q_{DNA} = 0.256$ Å^{-1} arises from DNA-DNA correlations and gives $d_{DNA} = 2\pi/q_{DNA} = 24.55$ Å (Fig. 2).

The precise nature of the packing structure of λ-DNA within the lipid layers can be elucidated by conducting a lipid dilution experiment in the isoelectric point regime of the complex. In these experiments the total lipid (L = DOTAP + DOPC) is increased while the charge of the overall complex, given by the ratio of cationic DOTAP to DNA, is kept constant at the isoelectric point DOTAP/DNA = 2.20. The SAXS scans in Fig. 4, (arrows point to the DNA peak) show that $d_{DNA} = 2\pi/q_{DNA}$ increases with lipid dilution from 24.54 Å to 57.1 Å as Φ_{DOPC} increases from 0 to 0.75 (or equivalently increasing L/D between 2.2 and 8.8). The most compressed interaxial spacing of ≈ 24.55 Å at $\Phi_{DOPC} = 0$ approaches the short-range repulsive hard-core interaction of the B-DNA rods containing a hydration layer. Fig. 5A plots d and d_{DNA} as a function L/D

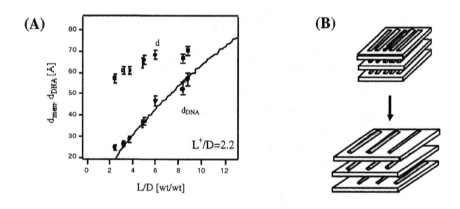

Fig. 5 (A) The DNA interaxial distance d_{DNA} and the interlayer distance d in the L_α^c phase (Fig. 2) plotted as a function of Lipid/DNA (L/D) (wt/wt) ratio at the isoelectric point of the complex DOTAP/DNA = 2.2. d_{DNA} is seen to expand from 24.5 Å to 57.1 Å. The solid line through the data is the prediction of a packing calculation where the DNA chains attached to the membrane form a space filling one-dimensional lattice. (B) Schematic of DNA-membrane multilayers showing the increase in distance between DNA chains as the membrane charge density is decreased (as Φ_{DOPC} increases) at the isoelectric point. (Adapted from references 5 and 29.)

The observed behavior is depicted schematically in Fig. 5B showing that as we add neutral lipid (at the isoelectric point) and therefore expand the total cationic surface we expect the DNA chains to also expand and increase their interaxial spacing. The solid line in Fig. 5A is derived from the simple geometric packing relationship $d_{DNA} = (A_D/\delta_m) (\rho_D/\rho_L) (L/D)$ which equates the cationic charge density (due to the mixture DOTAP$^+$ and DOPC) with the anionic charge density (due to DNA$^-$) and is only valid at the isoelectric point where there is no excess lipid or DNA coexisting with the complex. Here, $\rho_D = 1.7$ (g/cc) and $\rho_L = 1.07$ (g/cc) denote the densities of DNA and lipid respectively, δ_m the membrane thickness, and A_D the DNA area. $A_D = Wt(\lambda)/(\rho_D L(\lambda)) = 186$ Å^2, $Wt(\lambda) = $ weight of λ -DNA $= 31.5 *10^6/(6.022*10^{23})$ g and $L(\lambda) = $ contour length of λ -DNA $= 48502 * 3.4$ Å.

The agreement between the packing relationship (solid line) with the data over the measured interaxial distance from 24.55 Å to 57.1 Å (Fig. 5A) is quite remarkable given the fact that there are no adjustable parameters. The variation in the interlayer spacing d ($= \delta_w + \delta_m$) (Fig. 5A, open squares) arises from the increase in the membrane bilayer thickness δ_m as L/D increases (each DOPC molecule is about 4 Å to 6 Å longer than a DOTAP molecule). The observation, of a *variation in the DNA interaxial distance* as a function of the lipid to DNA (L/D) ratio in multilayers (Fig. 5A), unambiguously demonstrates that x-ray diffraction directly probes the DNA behavior in multilayer assemblies [5]. From the linewidths of the DNA peaks the 1D lattice of DNA chains is found to consist of domains extending to near 10 neighboring chains [7]. Thus, the DNA chains form a finite-sized 1D ordered array adsorbed between 2D membranes; that is, a finite sized 2D smectic phase of matter. On larger length scales the lattice would melt into a 2D nematic phase of chains due to dislocations.

The DNA-lipid condensation can be understood to occur as a result of release of "bound' counterions in solution. DNA in solution (Fig. 1) has a bare length between negative charges (phosphate groups) equal to $b_o = 1.7$ Å. This is substantially less than the Bjerrum length in water $bj = 7.1$ Å which corresponds to the distance where the Coulomb energy between two unit charges is equal to the thermal energy $k_B T$. A non-linear Poisson-Boltzmann analysis shows that counterions will condense on the DNA backbone until the Manning parameter $\xi = bj/b'$ approaches 1 [30]. (b' is the renormalized distance between negative charges after counterion condensation.) A similar analysis shows that counterions also condense near the surface of 2-dimensional membranes (i.e. within the Gouy-Chapman layer) [31]. Through DNA-lipid condensation the cationic lipid tends to fully neutralize the phosphate groups on the DNA in effect replacing and releasing the originally condensed counterions in solution. Thus, the driving force for higher-order-self-assembly is the release of counterions, which were one-dimensionally bound to DNA and two-dimensionally bound to cationic membranes, into solution.

3. The Role of the "Helper-Lipid" in controlling the Interaction Free Energy and Structures in Cationic Lipid – DNA Mixtures: Structural Transition From the lamellar L_α^C Phase to the Inverted Hexagonal H_{II}^C Phase.

The Helper Lipid is typically the neutral lipid that is mixed in with the cationic lipid before complexation with DNA. Thus, the helper lipid controls not only the charge density of the membrane but also the physical properties (such as the rigidity, spontaneous curvature) of the membrane. We expect that the interplay between the electrostatic and membrane elastic interactions in the complexes determine their structure. Recent theoretical work shows that electrostatic interactions alone are expected to favor the inverted hexagonal H_{II}^C phase (schematic in Fig. 6), which minimizes the charge separation between the anionic groups on the DNA chain and the cationic lipids [32,33].

$$H_{II}^C$$

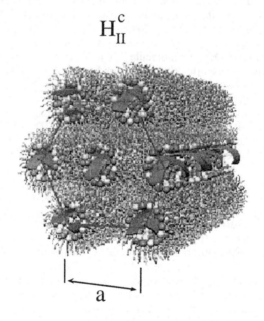

a

Fig. 6 Model of the inverted hexagonal H_{II}^C phase (cylinders consisting of DNA coated with a lipid monolayer arranged on a hexagonal lattice) of cationic lipid-DNA (CL-DNA) complexes. Model is based on synchrotron x-ray diffraction data (Adapted from reference 38)

The electrostatic interaction maybe be resisted by the membrane elastic cost of forming a cylindrical monolayer membrane around DNA. Indeed, lipids have a tendency to form a variety of shapes. For example, many lipids (e.g. phosphatidylcholine, phosphotidylserine, phosphotidylglycerol, cardiolipin) have a cylindrical shape, with the head group area $\approx$ the hydrophobic tail area, and tend to self-assemble into lamellar structures with a natural curvature $C_o = 1/R_o = 0$. Other lipids (e.g. phosphotidylethanolamine) have a cone shape, with a smaller head group area than tail area, and give rise to a negative natural curvature $C_o < 0$. Alternatively, lipids with a larger head group than tail area (some lysophospholipids) have $C_o > 0$.

3.1 CONTROL OF CL-DNA STRUCTURE BY VARYING SHAPES OF LIPID MOLECULES

It is well appreciated [34-37] that in many lipid systems the "shape" of the molecule which determines the natural curvature of the membrane $C_o = 1/R_o$ will also determine the actual curvature $C = 1/R$ which describes the structure of the lipid self-assembly (e.g. $C = 0 \rightarrow$ lamellar L_α; $C_o < 0 \rightarrow$ inverted hexagonal H_{II}; $C_o > 0 \rightarrow$ hexagonal H_I).

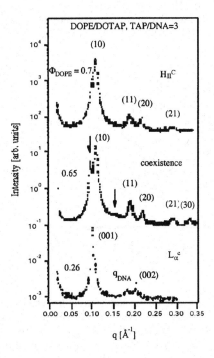

Fig. 7 Synchrotron SAXS patterns of the lamellar L_α^C and columnar inverted hexagonal H_{II}^C phases of positively charged CL-DNA complexes as a function of increasing weight fraction Φ_{DOPE}. At $\Phi_{DOPE} = 0.26$, the SAXS results from a single

phase with the lamellar L_α^C structure shown in Fig. 2. At $\Phi_{DOPE} = 0.7$, the SAXS scan results from a single phase with the columnar inverted hexagonal H_{II}^C structure shown in Fig. 6. At $\Phi_{DOPE} = 0.65$, the SAXS shows coexistence of the L_α^C (arrows) and H_{II}^C phases. (Adapted from reference 38)

This is particularly true if the bending rigidity of the membrane is large ($\kappa/k_BT \gg 1$), because then a significant deviation of C from C_o would cost too much elastic energy. However, if the bending cost is low with $\kappa \approx k_BT$, then C may deviate from C_o without costing much elastic energy especially if another energy is lowered in the process. Here we show experiments where the bending rigidity is large and the structure of the self assembly is controlled by the "shape" of the neutral lipid (the so-called Helper Lipid) used in the mixture of DNA and lipid.

We show in Fig. 7 SAXS scans in positively charged CL-DNA complexes for DOTAP/DNA (wt./wt.) = 3 as a function of increasing Φ_{DOPE} (weight fraction of DOPE). We find that the internal structure of the complex changes completely with increasing DOPE/DOTAP ratios [38]. SAXS data of complexes with $\Phi_{DOPE} = 0.26$ and 0.70 clearly shows the presence of two different structures. At $\Phi_{DOPE} = 0.26$, SAXS of the lamellar L_α^C complex shows sharp peaks at $q_{001} = 0.099$ Å^{-1} and $q_{002} = 0.198$ Å^{-1} resulting from the lamellar periodic structure ($d = 2\pi/q_{001} = 63.47$ Å) analogous to the structure in DOPC/DOTAP-DNA complexes (Fig. 2).

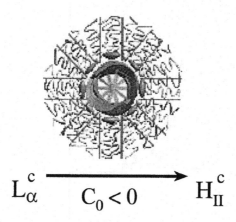

$$L_\alpha^c \xrightarrow{\quad C_0 < 0 \quad} H_{II}^c$$

Fig. 8 The H_{II}^C phase is expected to be the stable structure when the natural curvature (C_o) of the cationic lipid monolayer is driven negative by the addition of the helper-lipid DOPE. This is shown schematically where the cationic lipid DOTAP is cylindrically shaped while DOPE is cone-like with negative natural radius of curvature.

For $0.7 < \Phi_{DOPE} < 0.85$ the peaks of the SAXS scans of the CL-DNA complexes are indexed perfectly on a two-dimensional (2D) hexagonal lattice with a unit cell spacing of a = $4\pi/[(3)^{0.5}q_{10}] = 67.4$ Å for $\Phi_{DOPE} = 0.7$. Fig. 7 at $\Phi_{DOPE} = 0.7$ shows the first four order Bragg peaks of this hexagonal structure at $q_{10} = 0.107$ Å^{-1}, $q_{11} = 0.185$ Å^{-1}, $q_{20} = 0.214$ Å^{-1}, and $q_{21} = 0.283$ Å^{-1}.

The structure is consistent with a 2D columnar inverted hexagonal structure (Fig. 6), which we refer to as the H_{II}^C phase of CL-DNA complexes [38]. The DNA molecules are surrounded by a lipid monolayer with the DNA/lipid inverted cylindrical micelles arranged on a hexagonal lattice. The structure resembles that of the inverted hexagonal H_{II} phase of pure DOPE in excess water [35,36], with the water space inside the lipid micelle filled by DNA. For $\Phi_{DOPE} = 0.65$ the L_α^C and H_{II}^C structures coexist as shown in Fig. 7 (arrows point to the (001) and q_{DNA} peaks of the L_α^C phase) and are nearly epitaxially matched with a $\approx$ d. For $\Phi_{DOPE} > 0.85$ the H_{II}^C phase coexists with the H_{II} phase of pure DOPE which has peaks at $q_{10} = 0.0975$ Å^{-1}, $q_{11} = 0.169$ Å^{-1}, $q_{20} = 0.195$ Å^{-1} with a unit cell spacing of a = 74.41 Å.

We can understand the L_α^C to H_{II}^C transition as a function of increasing Φ_{DOPE} (Fig. 7) by noting that in contrast to the helper-lipid DOPC and the cationic lipid DOTAP (Fig. 2) which have a zero natural curvature ($C_o^{DOTAP,DOPC} = 1/R_o^{DOTAP,DOPC} = 0$), DOPE is cone-shaped with $C_o^{DOPE} = 1/R_o^{DOPE} < 0$. Thus, the natural curvature of the monolayer mixture of DOTAP and DOPE is driven negative with $C_o = 1/R_o = \Phi_{DOPE}C_o^{DOPE}$. Hence, as a function of increasing Φ_{DOPE} we expect a transition to the H_{II}^C phase favored by the elastic free energy as the data indicate. Thus, the helper-lipid DOPE appears to induce the L_α^C to H_{II}^C transition by controlling the spontaneous radius of curvature R_o of the lipid layers as shown schematically in Fig. 8.

4. Attractions between equally charged Bio-Polyelectrolytes adsorbed on Membranes: Condensation Transitions in 2-dimensions in the presence of Multivalent Counterions

The L_α^C multilayers where DNA is adsorbed onto membranes (Fig. 2) with very weak coupling between DNA chains from layer to layer [5,7,29] constitute a simple model system for studies directed towards understanding polyelectrolyte (e.g. DNA, F-actin, microtubule) condensation in two-dimensions. The mechanisms of DNA condensation in vivo are poorly understood. For example, in bacteria, it is thought that the polyamine molecules, which occur in abundant quantities are most probably responsible for DNA condensation. Simple model experiments in vitro have found that in the presence of multivalent cations with charge 3$^+$ or larger (e.g. the polyamines spermidine (3$^+$), spermine (4$^+$)) DNA condenses into compact objects with dimensions of approximately 50 nm [1]. Thus, it appears that DNA condensation in vivo and in vitro is controlled by multivalent cations, polyamines, peptides and proteins (e.g. histones in eukaryotic cells), and that electrostatic effects are important.

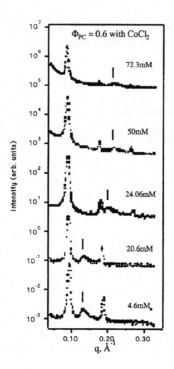

Fig. 9 Synchrotron SAXS patterns of the lamellar L_a^C phase of CL-DNA complexes as a function of increasing molarity M of Co^{2+}. Between M= 0.4.6 mM and $\approx$ 20.6 mM, the SAXS data show a constant spacing $d_{DNA} = q_{DNA}/2\pi = 48$ Å which matches the membrane charge density as described in C.3. Above M = 25 mM, the spacing undergoes an abrupt decrease to a condensed DNA-DNA interchain spacing of $d_{DNA} = 28$ Å (Adapted from reference 16)

From a fundamental physics view point, current experimental data are not in agreement with theory regarding the nature of Coulomb-induced attractive interactions between polyelectrolyte rods with the same sign in 3-dimensional solution. For example, experiments show that DNA forms condensed bundles only in the presence of multivalent counterions with valence equal or larger than 3^+ [1]. On the other hand, F-actin, which has about the same bare length (i.e. 2.2 Å between 2 negative charges instead of 1.7 Å for DNA) and a diameter (8.5 nm) near four times larger than DNA condenses in the presence of divalent 2^+ and larger valence cations [39]. Thus, experiments suggest that the attractive interactions are larger for F-actin. The effective interaction between polyelectrolyte rods embedded in a solution of their counterions was studied numerically by Brownian particle dynamics by Gronbech-Jensen et al. [40]. A significant range of attraction between the rods was identified as arising from the microscopic ordering of counterions in the vicinity of the polyelectrolytes. However, the MD study is not able to account for the difference in the behavior of polyelectrolyte condensation between DNA and F-actin, which appear to be simple model systems.

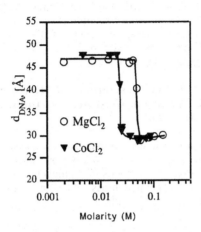

FIGURE 10. The DNA interchain spacing in L_α^C complexes at the isoelectric point (see Fig. 2) as a function of increasing divalent counterions. The membrane consists of DOPC/DOTAP mixtures with 60% DOPC which sets the decondensed inter-chain spacing around 48 Å as described in section 2. The DNA condensation transition (where the divalent cation electrostatically links neighboring DNA chains) is abrupt and occurs at lower concentrations for Co^{2+} which has a larger atomic number than Mg^{2+}. (Adapted from reference 16)

We have studied polyelectrolyte condensation in the lamellar L_α^C phase of CL-DNA complexes in the presence of multivalent cations [16]. As we described above, because we are able to precisely measure the DNA-DNA (and in principle other polyelectrolyte biopolymers) inter-chain distance with x-ray diffraction, the experiments allow us to clearly and quantitatively probe the behavior of polyelectrolyte chains in 2D revealing the various phases and intermolecular interactions. Data on L_α^C complexes in the presence of divalent counterions gives clear evidence of a short range attractive interaction between DNA chains, which is not observed in DNA chains suspended in 3-dimensional chains. The SAXS scans in Fig. 9, (vertical lines point to the DNA peak) show that $d_{DNA} = 2\pi/q_{DNA}$ is near constant for L_α^C complexes in the presence of Co^{2+} for molarity less than about 20 mM beyond which we see an abrupt decrease in d_{DNA}. Fig. 10 plots the results of these synchrotron experiments with L_α^C complexes in the presence of Mg^{2+} and Co^{2+}. We see that the DNA lattice adsorbed on membranes undergoes what appears to be a novel collective condensation transition with the DNA inter-chain spacing abruptly decreasing from 48 Å (the de-condensed repulsive state of DNA-DNA interactions described in section 2) to the collapsed state of 28 Å where attractive interactions between DNA chains dominates. Since the DNA diameter is 20 Å, the remaining 8 Å is just the size of hydrated Mg^{2+} and Co^{2+}. The intercalation of the multivalent counterions between DNA chains in the L_α^C complexes can be measured directly. The depletion of Co^{2+} from the supernatant solution surrounding the complex can be readily measured because Co^{2+} has a very strong absorbtion peak in the visible at 511 nm [16]. This enables one to quantitatively measure the Co^{2+} concentration above and below the condensation transition $\approx$ 21 mM. We have found that in the condensed phase (M > 21 mM), about 70% of the negative phosphate groups of DNA are "bound" with Co^{2+}. Thus, the condensation transition corresponds to the multivalent cations electrostatically cross-linking neighboring DNA chains.

5. Biomolecular Materials Self-Assembled on Micro- and Nano- Scale Patterned Surfaces: Orientation, Confinement-Induced new Phases, and Technological Applications

To obtain oriented phases and confinement-induced new phases consisting of the membranes complexed with polyelectrolyte biopolymers described in the previous sections we have recently developed methods for producing and characterizing the structures in such biomolecular materials self-assembled on patterned (linear channels, both varying and constant channel width) surfaces.

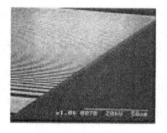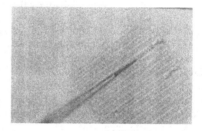

Fig.11 Left: Varying width linear microchannels fabricated on Si (111) substrates with channel width varying from 5 μm down to 500 nm. The microchannel's surfaces and confining dimensions control the orientation of self assembling biomolecules. Right: A constant width microchannel fabricated on glass with 10 μm channels and 1.5 μm deep. A micropipette with a tip size of order 1 μm is shown injecting sequential DNA then lipid solution into the channels. The in-situ formation of the complex can be seen in two filled channels. (Adapted from 41-43)

These patterned surfaces are produced by our group (in particular with Dr. Youli Li) at UCSB using optical and electron-beam lithography and reactive ion etching methods for fabrication of 10s of microns to about 0.1 micron features [41-43]. The microchannels are produced at the UCSB branch of the National Nanofabrication Users Network. Scanning electron microscopy is used to characterize and monitor for imperfections.

We have fabricated linear width microchannels with widths ranging between 20 μm down to 100 nm on a variety of substrates (Si, GaAs, (100 μm thick wafers) and glass plates (170 µm thick)). Channel depths may be as small as 2 nm (i.e. the etch resolution) and as large as five to ten times deeper than the width but in typical cases is about 1.5 µm. Fig. 11 shows examples of both a variable width linear microchannel (Left) and a constant width linear microchannel [41-43]. These are used for structural studies of biomolecules mixed in micron and sub-micron scale microchannel structures as substrate matrices for confining, orienting, and guiding the growth direction of the self assemblies. A variable width microchannel (e.g. Fig. 11 Left) can be used with a microbeam x-ray source with dimensions of order 1 µm^2. For example, initial demonstration experiments were recently done using the transmission zone plate spectrometer at the Advanced Photon Source to look for the effect of channel width on the orientation of the biomolecular self-assembly and also to readily find the optimal channel width, which gives a desired orientation.

Using recently developed methods the patterened surfaces are typically modified so that the valleys (grooves) are hydrophilic while the top surfaces are hydrophobic [41-43]. In this manner the charged membrane-polyelectrolyte biopolymer complex remains inside the grooves. Typically the entire pattern is initially rendered hydrophilic by a standard RCA method using a mixture of 30% H_2O_2, 30% NH^{3+}, H_2O (1:1: 5) which leaves both hydroxyl and O$^-$ groups on the surface. Next a 37% HCl and water mixture (1:1) converts most of the O$^-$ groups into hydroxyl groups. The next step modifies the pattern to render the top surfaces hydrophobic. Planar PDMS-stamps are used for the microcontact printing. For the formation of a hydrophobic self-assembled monolayer (SAM) on top of a silicon and glass surface OTS (Octadecyltrichlorosilane) is used and for a GaAs-surface alkylthiol is used. The SAM renders the top surfaces hydrophobic while the grooves are hydrophilic (negatively charged). We can also prepare positively charged grooves by the additional dipping of the patterned wafer in a PEI (Polyethyleneimine) water solution which changes the surface charges in the grooves. This then leads to a thin adsorbed hydrophilic PEI-layer with a positive charge.

In recent experiments we found significant improvement in the orientational quality of cationic lipid-DNA self asssemblies just after loading in the microchannels using a micropipette injector even before any temperature annealing had occured. The biopolymers, which can be explored, include DNA, F-actin, and microtubules. We expect the persistence length to be a critical parameter in achieving oriented biopolymers adsorbed onto membranes. The surfactant bilayers are expected to be oriented by the influence of the bottom surface and temperature annealing independent of the width of the channels. These experiments should lead to the possiblity of observing new phases and, in particular, a novel "confinement induced" sliding columnar phase predicted by recent theory where the biopolymers are orientationally locked (but positionally disordered) from layer to layer [13,14]. As we described above x-ray line-shape studies in unoriented multilayers of CL-DNA complexes allowed us to measure the distance-dependence of the 1D DNA lattice compressibility $B(d_{DNA})$ [7]. Future x-ray studies in aligned samples using microchannels will allow an independent measurement of κ_c and will open up a novel new way to determine the persistence length of surface adsorbed polyelectrolytes.

Aside from the structural studies, the oriented multilamellar structure would have many important technological applications. These oriented bioploymer/multilamellar structures confined in microchannels may be used to produce nanoscale materials. In principle, the anionic biopolymer (DNA, microtubule, F-actin) may be used as a template for adsorbing divalent cations on the route to producing nano-wires. We saw in the previous section 4 that in experiments on unoriented bulk samples we have found that in the presence of divalent cations (e.g. Mg^{2+}, Co^{2+}) the DNA lattice adsorbed on membranes undergoes a condensation transition with the DNA interchain spacing abruptly collapsing to 28 Å and the divalent cations electrostatically bound to and trapped between neighboring DNA chains (Fig. 10). Once the divalent cations are bound to the DNA template, two different reaction routes can be followed to form the desired chemical reactions leading to nanoscale wires with conducting or semiconducting properties. The first involves saturation in a chalcongenide gas (e.g. H_2S), and the second involves a reduction reaction (basic hydroquinone/OH⁻ solution) followed by an acidic reaction of hydroquinone/H^+ and cations [44].

We may also explore applications based on these oriented self assemblies in separations technologies. Here, the oriented self-assemblies will be used as templates for the synthesis of inorganic nanoporous materials using cationic inorganic molecular species as intercalating species in the nanoscale pore region (i.e. the aqueous space between the biopolymers DNA, F-actin, microtubules) [45].

6. Cell Cytoskeletal Assembly In Vitro

In this last part of the lectures we will give a brief discussion of the collective interactions between different proteins that lead to supramolecular assemblies, for example, of bundles of filamentous actin. Our long-term goal at UCSB is to understand the formation of structures, spanning lengths from the nanometer to the micrometer scale, in three distinct types of negatively charged filamentous proteins. These include, filamentous actin (F-actin), 8.5 nm in diameter, the intermediate filaments (IFs), 10 nm in diameter, and microtubules (MTs), 25 nm in diameter, which constitute the cell cytoskeleton (Fig. 12, left) [46-48]. Recent papers from our laboratory concerning new types of assemblies in F-actin and DNA can be found elsewhere [49-51].

Our primary focus is on self-assembly with biological rather than synthetic (e.g. block copolymers) macromolecules. One motivation is that we are still far from having a comprehensive understanding of the critical conditions required for self assembly between like charged polyelectrolytes, which typically exhibit purely repulsive electrostatic interactions [40,52,53]. Thus, it is not clear which synthetic polyelectrolyte will be a suitable model system for exploring these rarely occurring attractive forces. Nature on the other hand, not only routinely turns on attractive forces to assemble complexes from like-charged polyelectrolytes, but also manages to assemble different structures using the same polyelectrolyte. The different structures relate to different functions. For example, filamentous-actin is thought to self assemble to form either a tightly packed bundle phase, or loosely packed two-dimensional and three-dimensional gel network structures in eukaryotic cells. The distinct functions, resulting from these highly regulated structures, include cell motility, cell attachment, cell cytokinesis (the physical splitting of a cell into two daughter cells), and others. The self-assemblies in cytoskeletal filaments in vivo are mediated in most cases by complicated oppositely charged protein based linking molecules. Our strategy is then a straightforward one. In our experiments we use the same building blocks, namely cytoskeletal filaments, but study both simple model linkers (e.g. small multivalent counter-ions and synthetic peptides), and also the more complex cross-linking-proteins which attach filaments through non-covalent bonds, (most likely, a combination of electrostatic and short range hydrophobic interactions). The studies with model linkers allow us to clarify the nature of the interactions between filament and linker. Thus, our studies should impact our understanding of polyelectrolyte physics, a very important field of soft condensed matter, which remains poorly understood.

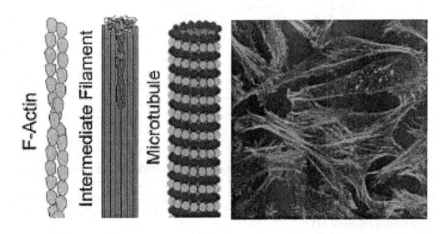

Fig. 12 Left: Cytoskeletal filaments F-actin (dia. ≈ 8.5 nm), Intermediate Filament (IF) (dia. ≈ 10 nm), Microtubule (MT) (dia. = 25 nm) (adapted from reference 46). F-actin and microtubules result from the self assembly of G-actin (globular units) and tubulin dimers (comprised of α-tubulin and β-tubulin shown in two colors) respectively. Intermediate filament assembly is more complicated (see e.g. reference 48). Right: 3-dimensional confocal microscope image of mouse fibroblast L-cells stained (Alexa-12379 green emission) to view the actin cytoskeleton (taken in the Safinya lab. by Dr. Alison Lin). Note stress fibers, microspikes, and radial bundles of filopodia in membrane-protrusions. The IF and MT also form interacting networks.

We show schematically in Fig. 12 (left) the cytoskeletal filaments, filamentous actin (F-actin), Intermediate filaments (IFs), and microtubules (MTs) with varying physical and chemical properties. The diameters of the biological polyelectrolytes increases from ≈ 85 Å for F-actin, to ≈ 100 Å for IFs, to 250 Å for microtubules. While F-actin, IFs, and MTs, are overall negatively charged, they contain both positive and negative charged groups. The persistence lengths cover a wide range from ≈ 1000 nm for IFs, to ≈ 3 to 10 μm for F-actin, to ≈ 1 mm for microtubules. F-actin is purchased from Cytoskeleton, Inc. Neurofilaments (the IFs of nerve cells with a branched structure) and tubulin are purified in Safinya's laboratory following methods developed by Liem [54] and further modifications made by graduate student Jayna Jones. Additionally, purified tubulin is obtained from Professor Leslie Wilson, our UCSB collaborator from the Biology department [55].

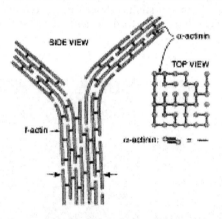

Fig. 13. A schematic of a possible model of F-actin bundles in the presence of the actin cross-linking protein α-actinin. The side view (only depicting a single layer from the bundle) shows a single branch (bifurcation) point. The top view shows a tentative square lattice which is highly disordered due to the absence of many linkers. (adapted from reference 51)

We show in Fig. 12 (right) laser scanning confocal microscope images of the actin cytoskeleton in mouse fibroblast cells. (The image is showing only the section closest to the cover slip to which the cell is attached.) The actin cytoskeleton forms different self-assemblies intended for distinct functions through actin cross-linking proteins. F-actin filaments can self-assemble into highly condensed bundles consisting of closely packed parallel filaments. Axial bundles form so-called stress fibers (Fig. 12, right), which end in focal adhesion spots and are responsible for cell adhesion. Radial bundles of filopodia in membrane-protrusions (Fig. 12, right) are also associated with adhesion complexes. F-actin filaments may associate with cross-linking proteins to form disordered gel-like networks for mechanical stability of the cell cytoplasm and cell shape, or attach to the plasma membrane forming a network for regulation of the distribution of membrane-proteins.

We show a schematic of a model of the interior of a bundle of F-actin cross-linked with α-actinin (shown at a branch point in Fig. 13). This model is based on recent synchrotron x-ray data combined with confocal microscopy, which shows that F-actin mixed with α-actinin, an actin cross-linking protein, forms what we call a "network of soft F-actin bundles" [51]. This network should exhibit novel length-scale dependent elastic behavior. Much remains to be studied about the precise structural nature of the various F-actin supramolecular self-assemblies and the solution conditions which lead to the formation of different types of bundles, networks, and other new types of self-assemblies.

7. Future directions.

A final comment regarding the motivation for studying such biological assemblies derives from the inherent importance of understanding the biophysics of self-assembly of cytoskeletal filaments. With the unraveling of the human genome and the emerging (post-genomic) proteomics era, the biophysics community is now challenged to elucidate the structures and functions of a large number of interacting proteins. Our understanding of modern biophysics has to a significant degree depended on our knowledge of the structure of the relevant biological molecules, which provides direct information relating to function. Macromolecular crystallography, which usually gives the structure of single protein molecules with angstrom resolution, is considered crucial to determining the structure of typically isolated proteins (e.g. where the overall globular protein occupies a volume of 3 to 6 nanometer cubed). We believe that the elucidation of structure in self-assembling biological systems comprised of a very large number of interacting proteins will require interdisciplinary approaches and technologies. Specifically, the simultaneous application of multiple techniques probing length scales from sub-nanometers to many microns, will be required to elucidate the interactions which result in the supramolecular structures of interacting biomolecules.

Acknowledgements

We acknowledge the contributions of group members, in particular, Youli Li, N. Slack, A. Lin, A. Ahmad, H. Evans, K. Ewert, T. Pfohl, O. Peltier, L. Hirst, N. Bouxsein, M. A. Ojeda-Lopez, D. Needleman, J. Jones, U. Raviv, J. Raedler, I. Koltover, T. Salditt, G. Wong, U. Schulze, and H-W Schmidt. We have benefited over the years through discussions with R Bruinsma, T Lubensky, P Pincus, B Gelbart, and A Ben-Shaul. The work was supported by grant GM-59288 from the National Institute of General Medical Sciences of the National Institute of Health (NIH) specific to DNA-lipid gene delivery studies and the National Science Foundation DMR 0203755 specific to biomaterials studies. The synchrotron x-ray diffraction experiments were carried out at the Stanford Synchrotron Radiation Laboratory which is supported by the U.S. Department of Energy. The Materials Research Laboratory at UCSB is supported by NSF-DMR-0080034.

48

8. References

1. See e.g. V. A. Bloomfield, "Condensation of DNA by Multivalent Cations - Considerations on Mechanism", *Biopolymers* **31**, 1471-1481 (1991).

2. See e.g. F. Livolant and A. Leforestier, "Condensed phases of DNA: Structures and phase transitions", *Prog. Poly. Sci.* **21**, 1115-1164 (1996).

3. J. Gustafsson, G. Arvidson, G. Karlsson, M. Almgren, "Complexes between Cationic Liposomes and DNA Visualized by Cryo-TEM", *Biochim. Biophys. Acta-Biomem.* **1235**, 305-312 (1995).

4. B. Sternberg, F. L. Sorgi, L Huang, "New Structures in Complex Formation between DNA and Cationic Liposomes Visualized by Freeze-Fracture Electron Microscopy", *FEBS letters* **356**, 361-366 (1994).

5. J. O. Raedler, I. Koltover, T. Salditt, C. R. Safinya, "Structure of DNA-Cationic Liposome Complexes: DNA Intercalation in Multi-Lamellar Membranes in Distinct Interhelical Packing Regimes", *Science* **275**, 810-813 (1997).

6. D. D. Lasic, H. H. Strey, M. C. A. Stuart, R. Podgornik, P. M. Frederik, "The structure of DNA-liposome complexes", *J. Am. Chem. Soc.* **119**, 832-833 (1997).

7. T. Salditt, I. Koltover, J.O. Raedler, C. R. Safinya, "Two Dimensional Smectic Ordering of Linear DNA Chains in Self-Assembled DNA-Cationic Liposome Mixtures", *Physical Review Letters* **79**, 2582-2585 (1997); T. Salditt, I. Koltover, J.O. Raedler, C. R. Safinya, "Self-assembled DNA-cationic Lipid Complexes: Two-Dimensional Smectic Ordering, Correlations, and Interactions", *Physical Review E* **58**, 889-904 (1998).

8. S. May, A. Ben-Shaul, "DNA-lipid complexes: Stability of honeycomb-like and spaghetti-like structures", *Biophysical J.* **73**, 2427-2440 (1997).

9. N. Dan, "The structure of DNA complexes with cationic liposomes - cylindrical or flat bilayers?", *Biochim.Biophys.Acta* **1369**, 34-38 (1998)

10. R. Bruinsma, "Electrostatics of DNA cationic lipid complexes: isoelectric instability", *European Phys. J. B* **4**,75-88 (1998).

11. R. Bruinsma and J. Mashl, "Long-range electrostatic interaction in DNA cationic lipid complexes", *Europhysics Letters*, **41** (2) 165-170 (1998).

12. D. Harries, S. May, W. M. Gelbart, A. Ben-Shaul, "Structure, stability, and thermodynamics of lamellar DNA-lipid complexes", *Biophysical J.* **75**, 159-173 (1998).

13. C. S. O'Hern and T. C. Lubensky, "Sliding columnar phase of DNA lipid complexes", *Physical Review Letters* **80**, 4345-4348 (1998).

14. L. Golubovic and M. Golubovic, "Fluctuations of quasi-two-dimensional smectics intercalated between membranes in multilamellar phases of DNA cationic lipid complexes", *Physical Review Letters* **80**, 4341-4344 (1998).

15. B. Lewin, B: *Genes VI*. Oxford: Oxford University Press; 1997.

16. I. Koltover, K. Wagner, and C. R. Safinya, "DNA Condensation in Two-Dimensions", *Proceedings of the National Academy of Sciences USA* **97** (26) 14046-14052 (2000).

17. G. C. L. Wong, Y. Li, I. Koltover, C. R. Safinya, Z. Cai, W. Yun, "Mesoscopic Structure of DNA-Membrane Self Assemblies: Microdiffraction and Manipulation on Lithographic Substrates", *Applied Physics Letters* **73** (14), 2042-2044 (1998).

18. Y. Li, G. C. L. Wong, E. J. Caine, E. L. Hu, C. R. Safinya, "Structural Studies of DNA-Cationic Lipid Complexes Confined in Lithographically Patterened Microchannel Arrays", *International Journal of Thermophysics* **19** (4), 1165-1174 (1998).

19. R. I. Mahato, S. W. Kim, editors "Pharmaceutical perspectives of Nucleic Acid-Based Therapeutics", (London: Taylor & Francis, 2002)

20. See e. g. A. D. Miller, "Cationic Liposomes for Gene Therapy" *Ang. Chem. (International Edition), Reviews* **37**, 1768-1785 (1998).

21. Alison Lin, Nelle Slack, Ayesha Ahmad, Cyril George, Chuck Samuel, C. R. Safinya, "Three-dimensional Imaging of Lipid Gene-Carriers: Membrane Charge Density Controls Universal Transfection Behavior in Lamellar Cationic Liposome-DNA Complexes", *Biophysical J.* **84**, 1-10 (2003).

22. K. Ewert, A. Ahmad, H. M. Evans, H. W. Schmidt, C. R. Safinya, "Efficient synthesis and cell-transfection properties of a new multivalent cationic lipid for nonviral gene therapy", *J. Medicinal Chemistry* **45**, 5023-5029 (2002).

23. See e.g. C. M. Henry, "Gene Delivery –Without Viruses", *Chemical & Engineering News*, (Cover Story) **79**, 35-41 (2001).

24. See e.g. D. Ferber, "GENE THERAPY: Safer and Virus-Free?" *Science,* **294**, 1638-1642 (2001).

25. See e.g. J. Alper, "Breaching the Membrane", in *Science, News Focus: Drug Delivery*, **296**, 838-839 (2002).

26. See e. g. S. Chesnoy, and L. Huang, "Structure and Function of Lipid-DNA Complexes for Gene Delivery", *Annu. Rev. Biophys. Biomol. Struct.*, **29**, 27-47 (2000).

27. C. R. Safinya, "Structures of Lipid-DNA Complexes: Supramolecular Assembly and Gene Delivery", *Current Opinion in Structural Biology* **11** (4) 440-448 (2001).

28. P. L. Felgner, T.R. Gadek, M. Holm, R. Roman, H.W. Chan, M. Wenz, J.R. Northrop, G.M. Ringold, and M. Danielsen, "Lipofection: A highly efficient, lipid-mediated DNA-transfection procedure", *Proc. Natl. Acad. Sci. USA* **84**, 7413-7417 (1987).

29. I. Koltover, T. Salditt, and C. R. Safinya "Phase Diagram, Stability and Overcharging of Lamellar Cationic Lipid - DNA Self Assembled Complexes", *Biophysical J.* **77** (2) 915-924 (1999).

30. G.S. Manning, "Limiting laws and counterion condensation in polyelectrolyte solutions. I. Colligative properties", *J. Chem. Phys.* **51**, 924-933 (1969)

31. B.H. Zimm and M. Le Bret, "Counter-Ion Condensation and System Dimensionality", J. *Biomolecular Structure and Dynamcis*, **1**, 461-471 (1983).

32. S. May, A. Ben-Shaul, "DNA-lipid complexes: Stability of honeycomb-like and spaghetti-like structures", *Biophysical J.* **73**, 2427-2440 (1997).

33. N. Dan, "The structure of DNA complexes with cationic liposomes - cylindrical or flat bilayers?", *Biochim.Biophys.Acta* **1369**, 34-38 (1998)

34. J. N. Israelachvili, *Intermolecular and Surface Forces* (Academic Press, London, second edition, 1992).

35. J. M. Seddon, "Structure of the Inverted Hexagonal (H_{II}) Phase, And Non-Lamellar Phase Transitions of Lipids", *Biochim.Biophys.Acta* **1031**, 1-69 (1989).

36. S.M. Gruner, "Stability of Lyotropic Phases with Curved Interfaces", *J. Phys. Chem.* **93**, 7562-7570 (1989).

37. M. J. Janiak, D. M. Small, G. G. Shipley, "Temperature and Compositional Dependence of the Structure of Hydrated Dimyristoyl Lecithin", *J. Biol. Chem.* **254**, 6068-6078 (1979).

38. I. Koltover, T. Salditt, J.O. Raedler, C. R. Safinya, "An Inverted Hexagonal Phase of DNA-Cationic Liposome Complexes Related to DNA Release and Delivery", *Science* **281**, 78-81 (1998).

39. J.X. Tang and P. A. Janmey, "The Polyelectrolyte Nature of F-Actin and the Mechanism of Actin Bundle Formation", *J. Bio. Chem.* **271**, 8556-8563 (1996).

40. N. Gronbech-Jensen, R. Mashl, R.F. Bruinsma, W.M. Gelbart, WM, "Counterion-induced attraction between rigid polyelectrolytes", *Phys. Rev. Lett.*, 1997 **78**, 2477-2480 (1997).

41. Y. Li, T. Pfohl, J. H. Kim, M. Yasa, Z. Wen, M. W. Kim, and C. R. Safinya, "Selective Surface Modification in Silicon Micro-fluidic Channels for Micro-manipulation of Biological Macromolecules", *Biomedical Microdevices* **3**(3) 239-244 (2001).

42. T. Pfohl, J. H. Kim, M. Yasa, H. P. Miller, G. C. L. Wong, F. Bringezu, Z. Wen, L. Wilson, Y. Li, M. W. Kim, and C. R. Safinya, "Controlled Modification of Microstructured Silicon Surfaces for Confinement of Biological Macromolecules and Liquid Crystals" *Langmuir* **17**, 5343-5351 (2001).

43. T. Pfohl, Y. Li, J. H. Kim, Z. Wen, G. C. L. Wong, M. W. Kim, I. Koltover, and C. R. Safinya, "Biological Polyelectrolyte-Cationic Membrane Complexes in Solution and Confined on Patterned Surfaces", *Colloids & Surfaces A: Physiochemical and Engineering Aspects* **198**, 613-623 (2002).

44. See e.g. P. V. Braun, P. Osenar, S. I. Stupp, "Semiconducting Superlattices Templated by Molecular Assemblies", *Nature* **380**, 325-328 (1996); J. L. Coffer, S.R., Bigham, X., Li, R.F., Pinizzotto, Y.G. Rho, R.M. PirHe, and I.L. PirHe, "Dictation of the Shape of Mesoscale Semiconductor Nanoparticle Assemblies by Plasmid DNA", *Appl. Phys. Lett.* **69**, 3851-3853 (1996).

45. Q. Huo, D. I. Margolese, U. Ciesla, D.G. Demuth, P. Feng, T.E. Gier, P. Sieger, A. Firouzi, B.F. Chmelka, F. Shuth, and G. Stucky, "Organization of Organic Molecules with Inorganic Molecular Species into Nanocomposite Biphase Arrays", *Chem. Mater.* **6**, 1176-1191 (1994).

46. H. Herrmann & U. Aebi, "Intermediate filaments & their associates:multitalented structural elements specifying cytoarchitechture and cytodynamics", *Curr. Opin. In Cell Biol.* **12**, 79-90 (2000).

47. P. A. Janmey, "The cytoskeleton and cell signaling: component localization and mechanical coupling", *Physiological Rev.* **78**, 763-781 (1998).

48. E. Fuchs & D. W. Cleveland, "A structural scaffolding of Intermediate Filaments in Health and Disease", *Science* **279**, 514-519 (1998).

49. "Supramolecular Assembly of Actin Filaments into a Lamellar Phase of Crosslinked Two-Dimensional Rafts", G. C. L. Wong, Alison Lin, Jay X. Tang, Youli Li, P. A. Janmey, C. R. Safinya, *Physical Review Letters* (published on-line 2 July 2003 in issue 1 of Volume 91, article 018103).

50. "DNA-Dendrimer Complexes: Quasi-2 Dimensional Columnar Mesophases", H. Evans, A. Ahmad, K. Ewert, T. Pfohl, A. Martin, R. F. Bruinsma, C. R. Safinya, *Physical Review Letters* (in press).

51. "Structure of Actin Cross-Linked with a-Actinin: A Network of Bundles", O. Pelltier, E. Pokidysheva, L. S. Hirst, N. Bouxsein, Y. Li, C. R. Safinya (submitted to *Physical Review Letters*).

52. F. Oosawa, *Polyelectrolytes* (Marcel Dekker, New York, 1971).

53. B.Y. Ha, A. J. Liu, "Counterion-mediated Attraction between 2 Like-Charged Rods". *Phys. Rev. Lett.* **79**, 1289-1292 (1997).

54. R. Liem, "Simultaneous Separation and Purification of Neurofilament and Glial filament proteins from brain", *J. Neurochemistry* **38**, 142-150, (1982).

55. W. B. Derry, L. Wilson, I. A. Khan, R. F. Luduena, & M. A. Jordan, "Taxol differentially modulates the dynamics of microtubules assembled from unfractionated and purified b-tubulin isotypes." *Biochemistry* **36**, 3554-3562, (1997).

SIMPLE EXAMPLES OF CELL MOTILITY

J.F.JOANNY
Physicochimie Curie
Institut Curie section Recherche, 26 rue d'Ulm 75248 Paris
Cedex 05, France

1. Introduction

Motility is a fundamental property of cells and bacteria [1, 2]. Unicellular organisms constantly move in search for food; eggs would not be fertilized in the absence of sperm cell motion; macrophages move to infection sites; fibroblast cells motion allows the remodeling of connective tissues and the rebuilding of damaged structures; cell motion also plays an important role in cancer with the formation of metastases. Cells and bacteria either swim when they are propelled by the motion of cilia or flagellae or by the polymerization of an actin gel, or they crawl on surfaces. The motion of crawling cells always involves a deformation of the cytoskeleton and it is usually done in several steps [3]: protrusion in which a new cytoskeleton structure polymerizes in front of the cell leading edge, adhesion of the cell on the substrate that allows momentum transfer to the substrate and depolymerization and contraction of the rear of the cell in which the adhesion sites are broken and the rear parts of the cytoplasm are dragged forward.

The cell cytoskeleton is a soft gel in general made of actin filaments [1]. The filaments are polar and have a well-defined plus end and a well-defined minus end. The elastic modulus of this gel is of the order of $10^3 - 10^4$Pa. The actin filaments are constantly polymerizing at the plus end, at the front of the moving cell and depolymerizing at their minus end towards the back of the cell. This phenomenon is known as treadmilling; it requires ATP and energy is thus consumed in the polymerization process. For most cells, the actin filaments of the cytoskeleton interact with myosin molecular motors [4]. Myosin molecular motors aggregate on the cytoskeleton gel and move towards the plus ends of the filaments. Their motion generates forces and also consumes energy in the form of ATP. If a motor aggregate is bound to two or more filaments, the action of the motors induces internal stresses in

A.T. Skjeltorp and A.V. Belushkin (eds.), Forces, Growth and Form in Soft Condensed Matter: At the Interface between Physics and Biology, 51-64.

the cytoskeleton gel. We call such a gel where energy is constantly injected and generates internal stresses an active gel. An active gel is by essence a structure which is not at thermal equilibrium.

We do not present here a general description of cell motility. We rather summarize recent results on very specific examples of cell motility that emphasize one of the general processes involved.

The bacteria *Listeria* swim and are propelled by the polymerization of an actin gel that forms a comet at the rear of the bacteria [7]. There are no myosin motors involved in this motion. The deformation of the gel creates elastic stresses that drive the motion. The polymerization kinetics is also coupled to the stress distribution, the polymerization velocity is large at places where the gel pulls on the bacteria and small at places where the gel pushes the bacteria (to create the motion). We present here a very short discussion of the motility of the bacteria following the work of Prost et al. [10, 9]. We then summarize our recent work on a biomimetic system where the bacteria are replaced by liquid oil drops: the deformation of the drops is a signature of the elastic normal stress distribution around the drop.

Nematode sperm cells do not swim as other sperm cells but they crawl [19, 20]. Their cytoskeleton is made of a protein called MSP which is very similar to actin but which is not polar. No molecular motors are involved in the motion of these cells. It has been suggested that the motion of nematode sperm cells is due to an interplay between internal stresses in the cytoskeleton due to a self-generated pH gradient and adhesion on the substrate. We give in section 3 a short account of our recent theoretical elastic model describing the competition between adhesion and cytoskeleton deformation;

The cytoskeleton of most cells is an active gel and cell motion involves the deformation of the gel under the action of the molecular motors. This has been studied experimentally on several types of cells such as fibroblasts or fish keratocyte cells[5, 6]. Keratocytes are our model system for cell motility because small cell fragments that contain mostly the active gel made of actin actin and myosins and which are therefore much simpler than the whole cell can move spontaneously. We present in section 4 a general hydrodynamic theory for active gels that could be the starting point of a model for cell motility.

2. Soft *Listeria*

2.1. *LISTERIA* MOTION

Listeria are bacteria with a sphero-cylinder shape of length L of the order of 10μm and radius R as sketched on figure 2.1. They are propelled by the polymerization of actin that forms a comet at the back of the bacteria [7].

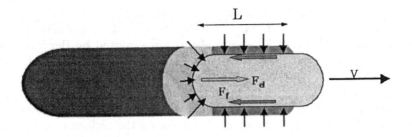

Figure 1. Sketch of listeria advancing under the elastic deformation of its actin comet

In a steady state, the velocity V is along the axis of the cylinder and we will here that the comet is a very long cylinder with a radius roughly equal to R. There is no actin gel in the front of the bacteria but the cylinder is coated with a thin gel layer of thickness $e \ll R$.

The motion of the bacteria results from a balance between the elastic force due to the elastic deformation of the gel and a friction force between the comet and the bacteria [10, 9]. The advancing velocity is of the order of $0.1\mu m/s$ and the hydrodynamic friction both on the gel and on the bacteria are completely negligible.

The friction force F_f is due to the attachment and detachment of the proteins that bind the gel to the bacteria. This force depends on the advancing velocity; we define a friction coefficient per unit area ξ and write the friction force as $F_f = 2\pi R L \xi V$. The friction constant depends on the advancing velocity. For simplicity, we assume here that it is constant. Some recent models have been proposed where the friction force varies non-monotonically with the velocity showing thus an unstable range of velocities [10]. This could explain the saltatory motion observed for mutants of *listeria* and for small latex beads propelled by actin polymerization [11].

Actin polymerization occurs at the surface of the bacteria under the influence of actin polymerization proteins. The polymerized actin is not under tension and on the surface of the bacteria the tensile stress σ_{tt} where t denotes the direction along the tangent of the bacteria surface vanishes. As new actin is polymerized on the surface, because of the curvature of the surface, the previously polymerized actin is put under tension and a tensile stress σ_{tt} builds up with a maximum value at the surface of the gel. On the surface, this stress is or order $\sigma_{tt} \sim Ee/R$ where E is the shear modulus of the gel. The elastic stress normal to the surface σ_{nn} is negative on the side surface of the cylinder as the gel is compressed due to the polymerization. It vanishes on the free surface of the gel. On the side of the cylinder, the thin shell approximation shows that the normal stress is of order $\sigma_{nn} \sim E(e/R)^2$. The total elastic force exerted by the comet on the bacteria is obtained by estimation of the elastic energy. It is of order $F_{el} \sim Ee^3/R$.

The elastic stress distribution on the surface of the bacteria affects the polymerization velocity. If for simplicity, we assume that the filaments are normal to the surface the force on a filament on the surface is $f = \sigma_{nn} a^2$ where a is the mesh size of the gel. The local polymerization velocity is then obtained from Kramers rate theory

$$v_p = v_p^0 \exp(\sigma_{nn} a^2 \delta/kT) \tag{1}$$

where T is temperature, k Boltzmann constant and δ a microscopic length of order of the actin monomer size. In a steady state, the polymerization velocity is much larger at the back of the bacteria where the polymerization occurs in the direction of the motion ($V = v_p$) than on the side where it occurs transverse to the motion. This imposes that the normal elastic stress is positive at the back of the bacteria thus pulling the bacteria backwards and negative in the vicinity of the contact line between the comet and the bacteria, thus pushing the bacteria.

A steady state is reached only if the time for polymerization at the back L/V is equal to the time for polymerization on the sides e/v_p where the polymerization velocity on the sides is approximately equal to the polymerization velocity in the absence of stress v_p^0. This fixes the gel thickness e and the force balance gives the advancing velocity [9]

$$V = v_p^0 \left(\frac{E}{\xi v_p^0}\right)^{3/4} (\frac{L}{R})^{1/2} \tag{2}$$

Current experiments are testing this law by measuring the advancing velocity under the action of an external force.

2.2. *LISTERIA* BIOMIMETICS: SOFT *LISTERIA*

In order to understand quantitatively the motion of *listeria* it is extremely useful to use biomimetic systems where objects with a well defined structure (that can be systematically varied) use the same actin propulsion mechanism as *listeria* [12, 14]. The difficulty is then the biochemistry of the system where all the necessary proteins must be introduced and the actin polymerization promoters must be linked to the surface of the object. This has been possible with spherical latex beads which have allowed a detailed study of the gel polymerization kinetics and of the saltatory motion due to a "negative" friction[11].

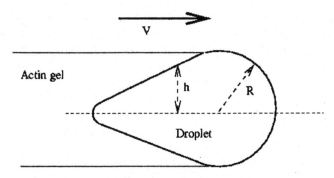

Figure 2. Deformation of an oil drop propelled by actin polymerization

Another spectacular result obtained on *listeria* is the change of the sign of the normal stress which is negative at the front corresponding to a compression of the gel and positive at the back corresponding to a stretching of the gel which pulls the bacteria backwards. This change of sign in the stress cannot be explained by molecular models that describe the motion in terms of the polymerization of single filaments at the back of the bacteria [8]. An adapted biomimetic system to study the stress distribution is a deformable object such as a liquid drop or a vesicle. Experiments have been performed both on liquid drops [17] and vesicles [13, 15, 16] and show a deformation of the spherical object to a pear-like shape with a spherical front part of radius R not covered by the actin gel and a cone-like rear part attached to the comet as shown on figure 2.2. The radius of curvature r at the back of the drop is much smaller than at the front.

We focus here on the experiments on liquid drops presented in reference [17]. We assume that we know the advancing velocity (experimentally of order $0.1 \mu m/s$ and we only focus on the shape of the drop. The advancing velocity could be obtained form a force balance argument similar to the one described in the previous paragraph. The shape of the drop is obtained from the balance of normal stresses at the surface of the drop. Applying Laplace's law both in the front part where there is no gel and in the back part where the drop in contact with the comet, we obtain

$$\frac{2\gamma}{R} = \gamma(\frac{1}{R_1} + \frac{1}{R_2}) - \sigma_{nn} \tag{3}$$

where R_1 and R_2 are the local curvature radii at any point in the cone region. We ignore here any change in the surface tension γ along the interface. This is far from being obvious as the actin polymerization promoters are present only at the back of the drop in the region where the drop is in contact with the comet. It has been argued however in reference [17] that they only give a very small contribution to the surface tension. We describe the shape of the drop by the variation of the angle θ between the local tangent to the drop and the direction of motion as a function of the liquid thickness h perpendicular to the motion. The total curvature is then $\frac{1}{R_1} + \frac{1}{R_2} = \frac{\cos\theta}{h} + \frac{d\cos\theta}{dh}$. The pressure balance (3) must be combined with the equation giving the polymerization velocity (1) and the volume conservation of the gel. If we assume that the comet is a perfect cylinder volume conservation is written as $v_p = V\sin\theta$.

Scaling the stress with the surface tension, we define a unit length $\ell = \frac{kT}{\gamma a^2 \delta}$ from equation (1) and a dimensionless number $\epsilon = \ell/R$. We discuss the shape of the drop in the limit $\epsilon \ll 1$ which corresponds to the experiments. For small values of ϵ the tangent angle θ is approximately constant such that $\sin\theta_0 = v_p^0/V$. At the very rear of the drop the tangent angle increases from this value to $\pi/2$ at the rear point over the small length ℓ. In the vicinity of the rear point, the shape of the drop is such that $h^2 = 4\ell z/\log(1/\sin\theta_0)$. The length ℓ gives thus the radius of curvature at the rear point $r = 2\ell/\log(1/\sin\theta_0)$.

The normal elastic stress distribution at lowest order in ϵ is

$$\sigma_{nn}(h) = \frac{\gamma}{R}\left(\frac{R\cos\theta_0}{h} - 2\right) \tag{4}$$

As expected the stress is large and positive at the rear of the drop and negative in the front part. It vanishes for a liquid thickness $h = h_{max}/2$ where $h_{max} = R\cos\theta_0$ is the liquid thickness at the triple line between the oil drop, the comet and the surrounding liquid. Equation 4 is not valid over a region of size ℓ in the vicinity of the rear point of the drop and the stress remains finite.

In reference [17] we also have solved numerically equations (3, 1) and compared systematically both the numerical and the approximate solutions to the shape of an experimental drop and and to the experimental stress distribution obtained from the experimental shape using equation (3). The agreement is very good with two adjustable parameters $\sin\theta_0 = 0.58$ and $\epsilon = 0.05$. This gives us a measure of two material parameters the length $\ell = 0.125\mu\text{m}$ and the polymerization velocity in the absence of stress $v_p^0 = 1.4\text{nm/s}$.

The calculated stress profile describes accurately the experimental stress profile except in a small region in the vicinity of the triple line between the external fluid, the oil droplet and the comet. At the triple line, the experimental stress does not vanish. In the calculation there is a discontinuity in the total curvature of the liquid drop: the curvature is $2/R$ in the spherical part of the drop and $1/R$ in the conical part. The elastic normal stress in the comet must therefore be equal to $-\gamma/R$ according to Laplace's law. This is inconsistent with the fact that the gel thickness vanishes at the triple line and thus cannot sustain any normal stress. We have solved this problem in reference [17] by introducing a boundary layer i.e. a small region in the vicinity of the triple line where the comet is not a perfect cylinder but bends towards the droplet. The elastic normal stress then decays smoothly to zero over this boundary region.

3. Nematode sperm cells

Nematode sperm cells have a cytoskeleton which is made of a globular protein called MSP. Quite like actin, this protein polymerizes and forms fibers that assemble to make a gel. The MSP protein is not however polar and does not interact with molecular motors. There exists a detailed experimental study of the motion of nematode sperm cells due to the group of Roberts and Stewart [24, 23, 25, 22, 26]. The moving cell forms a long and thin lamellipodium that drags the thicker cell body. The pH in the lamellipodium is not constant; the cell generates an internal pH gradient by injection of protons at the back of the lamellipodium. The polymerization and depolymerization of the MSP protein strongly depends on the local pH. At the leading edge of the cell where the pH is larger, MSP polymerizes; at the rear of the lamellipodium where the pH is lower MSP depolymerizes. In a steady state, the depolymerization exactly compensates the polymerization. pH also influences the swelling of the MSP cytoskeleton. The MSP gel is more contracted at the back where pH is lower. We describe the local contraction by factor $\Lambda \leq 1$ which gives the contraction of the distance between two points with respect to their distance in the polymerization state at the leading edge of the cell.

A model for the motion of nematod sperm cells has been proposed by Bottino et al. [18]. Upon polymerization of MSP at the front, the cytoskeleton adheres on the support and this generates internal stresses. The adhesion is modeled by a viscous friction that depends on the local pH; the viscous friction is lower at the back and higher at the front. The internal stress is also a given function of pH. The model requires thus a gradient in adhesion and a prescribed internal stress.

We have proposed in reference [21] a model for the motion of nematode sperm cells that requires neither an adhesion gradient nor a prescribed internal stress. For simplicity, our model is two-dimensional, with a horizontal direction x along the motion and a direction z transverse to the substrate (see figure 3). We describe the cytoskeleton as a thin gel layer of constant thickness h much smaller than its length L. The adhesion is due to localized adhesion points that rupture if the horizontal force acting on them is larger

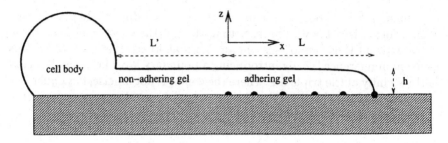

Figure 3. Advancing sperm cell on a substrate

than a critical value f_c. We consider here the critical force as a constant but it may depend weakly on the advancing velocity as predicted by classical theories. As the cytoskeleton polymerizes, it adheres on the substrate. In the reference frame of the cell, a given element of gel is transported by the cell motion towards the rear of the cell where the pH is lower and where it should thus become more contracted. The contraction is prevented by the adhesion of the surface; this generates internal elastic stresses on the substrate that eventually rupture the adhesion points. The cytoskeleton gel has thus two parts, an adhering part at the front carrying the internal stress and a non adhering part which is contracted due to the pH gradient but where the stress has relaxed.

Using the thin shell approximation, we have calculated in reference [21] the elastic shear stress on the substrate in the adhering part of the cytoskeleton. In most of the adhering part, the stress is negative and the gel is pulled backwards by the more contracted gel The thin shell approximation leads to

$$\sigma_{xz} = -2\mu\frac{\lambda + \mu}{\lambda + 2\mu}h\frac{d(\Lambda^2 - 1)}{dx} \tag{5}$$

where λ and μ are the two Lamé coefficients of the gel. The macroscopic force balance imposes that the total force on the cytoskeleton vanishes. The force is only due to the substrate as the viscous friction on the cell body is negligible this imposes that $\int dx\sigma_{xz} = 0$. In the vicinity of the last adhesion point where adhesion ruptures, over a region of size h the stress is thus positive and the gel is pulled forward. Assuming that the stress is constant in this region, we estimate it as

$$\sigma_{xz}(x = L) \sim \mu\frac{\lambda + \mu}{\lambda + 2\mu}(1 - \Lambda^2) \tag{6}$$

The stress is much larger at the back where it is positive and adhesion indeed ruptures at the back. At the rupture point, the stress is balanced by the adhesion force nf_c where n is the density of adhesion points. The stress given by 6 only depends on the contraction factor Λ, there is therefore a critical value of the contraction factor or of pH where adhesion ruptures. One of the experimental tests of our theory would be a measure of the tangential stress distribution on the substrate in order to test the change of sign at the back of the adhering cytoskeleton.

The advancing velocity of the nematode sperm cell is in our model fixed by the polymerization or the depolymerization velocity (which are equal in a steady state). It can be modified by applying an external force on the cell. The deformation of the gel is only very locally affected by an external force over a region of size h. An external force has thus no effect on the adhesion of the cytoskeleton except if the force is applied at the very rear of the adhering part of the cytoskeleton. If the force is applied at the rear of the cytoskeleton it effectively increases the critical rupture force for adhesion and the adhering part of the cytoskeleton is longer.

An external force also changes the polymerization and depolymerization velocities. A force opposing the motion applied at the leading edge of the cell decreases the polymerization velocity and tends to slow down the cell. More surprisingly, a force applied at the back of the cell increases the depolymerization velocity and thus the cell advancing velocity. Our model predicts thus a negative mobility. The effect of an external force would be a strong test of our theoretical approach.

4. Rotating vortices in active gels

4.1. HYDRODYNAMICS OF ACTIVE GELS

The cytoskeleton of most cells is an active gel made of actin filaments and myosin molecular motors. We have presented in reference [30] a generalized hydrodynamic theory of actin gels. The basic idea is to write down linear dynamic equations respecting all the symmetries of the problem for the relevant hydrodynamic variables. There are 4 components in an active gel. Actin can be in the monomeric form or polymerized in the gel. Actin monomers diffuse freely in the solution and they participate to the polymerization and depolymerization reactions. These reactions occur at the surface of the gel. As actin polymerizes and depolymerizes matter is injected into the actin gel (and extracted from the actin gel); this induces a flow of the actin gel. We describe this flow by the actin velocity field $\mathbf{v}$.

The actin filaments forming the gel are polar. We characterize the polarity by a polarization field $\mathbf{p}$ given by the local average of the unit vector pointing towards the plus end of all the filaments. There is a thermodynamic conjugate field to the polarization, $\mathbf{h}$. The field has two components a parallel component fixing the module of the polarization and a transverse component that creates a torque which aligns the polarization. In a two dimensional description where the polarization is parametrized by a polar angle ϕ, we use the standard theory of nematic liquid crystals and in the approximation where the two elastic constants are equal ($K_1 = K_3 = K$), the torque is $K\nabla^2\phi$ [28].

Myosins exist also under two forms, free myosins diffuse freely in the gel. Myosins also associate and bind to the gel. We describe this association as an nth order chemical reaction where n is a number which is found experimentally of order 3 or 4 [6]. The flux of bound myosins is convective. They are bound to the actin gel and they are transported by the flow of the actin gel. But myosins are molecular motors and they move relative to the actin filaments forming the gel in the direction of their plus ends. This motion requires the consumption of ATP; we characterize the ATP consumption by the chemical potential difference $\Delta\mu$ between ATP and its hydrolysis products. We make here a linear hydrodynamic theory and therefore linearize all variables with respect to $\Delta\mu$. The velocity of the myosin motors with respect to the actin filaments is written as $u\Delta\mu\mathbf{p}$ where u is a phenomenological constant. The symmetry of the problem imposes a myosin velocity along the polarization $\mathbf{p}$.

The linear generalized hydrodynamic theory of the active gel is constructed by analogy with the hydrodynamic theory of Martin Parodi and Pershan of liquid crystals [29]. We write a linear relation between fluxes and forces respecting all the symmetries of the problem, including the time reversal symmetry. In addition to the molecular fluxes of actin and myosins, we choose as fluxes the mechanical stress σ_{ij}, the time derivative of the polarization field $\mathbf{P} = \frac{d\mathbf{p}}{dt}$ and the ATP consumption rate r (the number of ATP molecules hydrolyzed per unit time). The conjugate variables are the velocity gradient u_{ij}, the field $\mathbf{h}$ and the ATP chemical potential $\Delta\mu$. The relation between fluxes and forces is a mobility matrix with two essential properties: the Onsager symmetry relations between off-diagonal mobility coefficients are satisfied and off-diagonal terms between fluxes having a different time reversal signature vanish. The active terms in the various fluxes are the terms proportional to $\Delta\mu$.

A complete derivation of the mobility matrix is given in references [30, 31]. As the expressions are rather lengthy, we only give here the expression of the mechanical stress. In the absence of ATP, the gel is described by a standard Maxwell model; it is liquid over long time scales and solid over short time scales with a single relaxation time τ. The mechanical stress is given by

$$2\eta u_{ij} = \left(1 + \tau \frac{D}{Dt}\right)\sigma_{ij} + \zeta\Delta\mu p_i p_j + \bar{\zeta}\Delta\mu\delta_{ij} - \frac{\nu_1}{2}(p_i h_j + p_j h_i) - \bar{\nu}_1 p_k h_k \delta_{ij} + \tau A_{ij}$$

$$(7)$$

η is the macroscopic viscosity of the cytoskeleton. There are 4 types of terms on the right hand side of equation 7. The first term is the standard convected Maxwell model where the convected derivative takes into account the geometrical non-linearities. The terms proportional to $\Delta\mu$ are the active components of the stress; they describe the internal stresses created by the myosins in the active gel. The two terms proportional to ν_1 and $\bar{\nu}_1$ describe the coupling between the stress and the polarization, they are standard terms in liquid crystal hydrodynamics. The last term A_{ij} summarizes the non linear couplings between the stress and the velocity gradients. These terms are present in the Oldroyd model for the rheology of viscoelastic fluids [27].

The final equation required to study the hydrodynamic of the active gel is the local force balance $\partial_j(\sigma_{ij} - P\delta_{ij}) - \xi v_i = 0$. P is the pressure that insures the incompressibility of the gel. We have included here a friction with the substrate with a local friction coefficient ξ.

4.2. SPIRAL DEFECT

An example of application of the generalized hydrodynamic theory given in reference [30] is that of a point-like defect of topological charge 1. In the keratocyte cell fragments that spontaneously move, the actin polarization is perpendicular to the edge of the cell. When the cell does not move it has the shape of a circular pancake on the substrate. If the symmetry is broken for example by a mechanical action the cell remains quasi-two-dimensional but it takes the shape of a bent peanut. There must then be defects of the polarization field. One possibility is that the cell fragment has two point-like defects on the sides. The counter-rotating motion of the two defects would then induce a motion of the cell as sketched on figure 4.2. Note however that the complete description of the motion would require a quantitative description of the adhesion on the substrate similar to what has been done for nematode sperm cells.

The simplest point like defect of two-dimensional nematics or XY magnets (corresponding to the polarization field) has a topological charge 1 [28]. It can be an aster where the polarization points radially to the center of the defect, a vortex where it is perpendicular to the radius from the center or a spiral defect where it makes a constant angle θ_0 with the radius. In this last case, the polarization field lines are spirals starting from the defect center. For a passive system, all defects have the same energy and

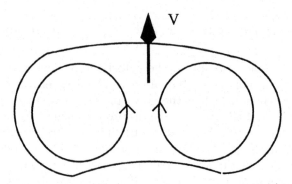

Figure 4. Possible mechanism for the motion of a keratocyte cell fragment due to the counter-rotation of two point-defects

are thus equally likely found if the two nematic elastic constants are equal. A larger splay constant favors the asters and a larger bend constant favors the vortex.

The hydrodynamic equations of the active gel are solved in reference [30] for a topological defect of charge 1. Even in the case where the two nematic elastic constants are equal, all defects are not equivalent and only four types of spiral defects are found at an angle θ_0 such that $\cos 2\theta_0 = 1/\nu_1$. The defect is thus stable only if $|\nu_1| \geq 1$; these angles are the same found for the orientation of a nematic liquid crystal in a shear gradient. The active spiral defect is associated to a velocity field in the orthoradial direction corresponding to a rotation of the defect. The velocity at a distance r from the center of the defect is

$$v_\theta(r) = 2\omega_0\lambda_f \left\{ K_1(\frac{r}{\lambda_f}) - \frac{\lambda_f}{r} \right\} \tag{8}$$

where $\omega_0 = \frac{2\sin 2\psi_0}{4\eta + \gamma_1\nu_1^2 \sin^2 2\psi_0}\tilde{\zeta}\Delta\mu$, γ_1 being the rotational viscosity of the gel and $\lambda_f = (4\eta + \gamma_1\nu_1 \sin^2 2\psi_0)^{1/2}/2\xi^{1/2}$. $K_1(x)$ is the modified Bessel function of order 1. The relative stabilities of the active and static defects is discussed in details in reference [30]. The essential result is that because of the active component of the stress (due to the molecular motors) polarization defects rotate in an active gel. This rotation could be at the origin of the motion of keratocyte cells as explained above.

5. Concluding remarks

We have presented in these lecture three example of cell motility where in each case a specific aspect is emphasized: actin polymerization for *listeria*, adhesion and its interplay with internal stresses for nematode sperm cells and the hydrodynamics of active actin-myosin gels for keratocyte cells. A more general theory of cell motility should include all these aspects. It does not seem to exist at the present time but the generalized hydrodynamic theory of active gels is a good starting point. Such a theory will however need precise and quantitative comparisons to experiments in order to measure all the phenomenological constants that must be introduced (for example in the mobility matrix). Our approach is a mesoscopic approach that ignores as much as possible the molecular details and the numerous proteins involved in the motility process. It could be completed by a more molecular description that would link the mesoscopic parameters to the molecular parameters such as the specificity of the various proteins in order to make a more direct link to biological experiments.

Aknowledgements The research presented here has been done in collaboration with J.Prost, O.Campas, H.Boukellal and C.Sykes (Institut Curie, Paris), F.Jülicher and K.Kruse (Dresden) and K.Sekimoto (Strasbourg).

References

1. Alberts B. et al. (2002) *Molecular biology of the cell*, Garland science, New York.
2. Bray D. (2000) *Cell movements*, Garland publishing, New York.
3. Mitchison T., Cramer L. (1996), *Cell* **84**, 371.
4. Howard J. (2001), *Mechanics of motor proteins and the cutoskeleton*, Sinauer Associates Inc., Sunderland.
5. Verkhovsky A., Svitkina T. and Borisy G. (1998), *Current Biology* **9**, 11.
6. Verkhovsky A., Svitkina T.and Borisy G., (1999), *Curr. Biol* **9**, 11.
7. Pollard T., Blanchoin L., Mullins R . (2000), *Annu Rev Biophys Biomol Struct.* **29**, 545.
8. Mogilner A, Oster G. (1996), *Biophys J.* **29**, 303 .
9. Gerbal F., Chaikin P., Rabin Y., Prost J. (2000), *Biophys J.* **79**, 2259.
10. Prost J. (2002), The physics of Listeria motion in *Physics of bio-molecules and cells*, Les Houches Lecture Notes LXXV, H.Flyvberg, F.Jülicher, P.Ormos and F.David eds. EDP Sciences, Springer.
11. Bernheim-Groswasser A., Wiesner S., Golsteyn R., Carlier M., Sykes C. (2002), *Nature* **417** 308.
12. Camerom L., Footer M., van Oudenaarden A., Theriot J.(2000), *Proc.Natl.Acad.Sci. USA* **96**, 4908.
13. Taunton J., Rowning B., Coughlin M., Wu M., Moon R., Mitchison T., Larabell C. (2000), *J.Cell Biol.* **148** 519.
14. Noireaux V., Golsteyn R.M., Friederich E., Prost J., Antony C., Louvard D., Sykes C. (2000), *Biophys J.* **78**, 1643.
15. Upadhyaya A., Chabot J., Andreeva A., Samadani A., and van Oudenaarden A. (2003), *Proc.Nat.Acad.Sci.* **100**, 4521.
16. Giardini P., Fletcher D., Theriot J. (2003), *Proc.Nat.Acad.Sci.* **100**, 6493.
17. Boukellal H., Campas O., Joanny J.F., Prost J., Sykes C. (2003), Soft listeria: actin based propulsion of liquid drops submitted to *Phys.Rev.Lett.*

18. Bottino D., Mogilner A., Roberts T., M.Stewart, Oster G. (2002), *J.Cell Sci.* **115**, 367.
19. Theriot J. (1996), *Cell* **84**, 1.
20. Roberts T., Stewart M. (2000), *J.Cell.Biol.* **149**, 7.
21. Joanny J.F., Jülicher F., Prost J. (2003), *Phys.Rev.Lett.* **90**, 168102.
22. King K., Stewart M., Roberts T. (1994), *J.Cell Sci.* **107**, 2941.
23. Italiano J., Roberts T., Stewart M., Fontana C. (1994), *Cell* **84**, 105.
24. Roberts T., Salmon E., Stewart M. (1998), *J.Cell Biol.* **140**, 367.
25. Italiano J., Stewart M., Roberts T. (1999), *J.Cell Biol.* **146**, 1087.
26. King,K., Essig J., Roberts T., Moerland T. (1994), *Cell Motil.Cytoskel.* **27**, 193.
27. Bird R. *et al.* (1987), *Dynamics of Polymeric Liquids* 2nd ed. Wiley, New York; Larson R. (1998), *Constitutive Equations for Polymer Melts and Solutions* Butterworth-Heinemann.
28. De Gennes P.G. and Prost J. (1993) *The Physics of Liquid Crystals* Clarendon Press, Oxford.
29. Martin P., Parodi O., Pershan P. (1972), *Phys.Rev.A* **6**, 2401.
30. Kruse K. et al. (2003), Asters, Vortices and Rotating Spirals in Active Gels of Polar Filaments submitted to *Phys.Rev.Lett.*
31. K. Kruse *et al* (2003), to be published.

STATISTICAL PHYSICS OF UNZIPPING DNA

David R. Nelson
Lyman Laboratory of Physics
Harvard University
Cambridge, MA 01238
nelson@cmt.harvard.edu

Abstract   The denaturation of double-stranded DNA as function of force and temperature is discussed. At room temperature, sequence heterogeneity dominates the physics of single molecule force-extension curves starting about 7 piconewtons of below a ~15 *pN* unzipping transition. The dynamics of the unzipping fork exhibits anomalous drift and diffusion in a similar range above this transition. Energy barriers near the transition scale as the square root of the genome size. Recent observations of jumps and plateaus in the unzipping of lambda phage DNA at constant force are consistent with these predictions.

Keywords: DNA, statistical mechanics, denaturation.

1. Single Molecule Biophysics Experiments

The past decade has seen a revolution in biophysics, due to exquisitely sensitive experiments [1] which probe ingredients of the "central dogma" of molecular biology [2] at the level of *individual* molecules. Although exceptions exist, the central dogma states that biological information first flows from DNA to messenger RNA via transcription mediated by a RNA polymerase. Information is then transferred from mRNA to proteins via a ribosome-mediated translation process.

Among the efforts to probe the basic constituents further: (a) Proteins anchored to a microscope slide in their biologically relevant folded state have been teased apart with atomic force probes [3] attached using, e.g., biotin-streptavidin linkages; (b) individual DNA's linked to magnetic beads have been stretched and twisted by small magnetic field gradients, allowing studies of both supercoiling and twist-induced denaturation [4]; and (c) the reversible unfolding of single RNA molecules (with beads attached to both ends) has been studied using laser tweezers

A.T. Skjeltorp and A.V. Belushkin (eds.), Forces, Growth and Form in Soft Condensed Matter: At the Interface between Physics and Biology, 65-92.

with a feedback loop to generate a constant force F [5]. In this later experiment, the fluctuating displacements of a ~25 base-pair RNA hairpin evolve from a predominantly closed state (with rare opening events) at small force to a predominantly open state (with rare closing events) at larger forces. At an intermediate force of $F_c \approx 14$ piconewtons (pN), the hairpin spends approximately equal time in both configurations, suggestive of a first-order phase transition between the open and closed states in the "thermodynamic limit" of a very large hairpin.

DNA force denaturation experiments related to the subject of this review have been carried by Bockelmann *et al.* [6–8] (see Fig. 1). In this pioneering work, the 48,502 base pairs of the virus phage lambda with one strand attached to a microscope slide were unzipped by a microneedle attached to the other strand under conditions which produced a constant rate of unzipping at a fixed temperature. Reasonable agreement with these experiments was obtained by direct numerical evaluation of the equilibrium statistical sums associated with a given average degree of unzipping. If the rate of unzipping is slow, such experiments effectively probe the physics at constant extension. The average force under these conditions locks in at about 14–15 piconewtons. The time-dependent fluctuations of this force as the unzipping proceeds provide information about the particular sequence of G:C and A:T base pairs being torn apart [6–8].

The full phase diagram of DNA in the force-temperature plane is shown in Fig. 2 [9–13]. The thermal DNA melting transition at zero force (see Sec. 2) is characterized by diverging length scales associated with denaturation bubbles. When $F \neq 0$, the native duplex DNA denatures into an unzipped state via a first-order phase transition (see Sec. 3). Under conditions of a slow, constant rate of extension [6–8], the average force *adjusts* so that the system sits on the heavy first-order transition line $F_c(T)$. At any stage in the unzipping process, the unzipping fork separating the native and unzipped states is like the meniscus dividing, say, liquid and gas phases at constant volume at a bulk first-order phase transition.

Although running unzipping experiments in a two-phase region associated with the line $F_c(T)$ is well-suited for possible DNA sequencing applications [6–8], the full force-temperature phase diagram of Fig. 2 is interesting for a number of reasons. First, understanding DNA denaturation under conditions of constant force (instead of constant extension) could provide insights into the complicated process by which DNA replicates during bacterial cell division [15]. Second, statistical mechanics in ensembles with *intensive* variables like the force held fixed is often more tractable than when the control parameter is a conjugate variable like

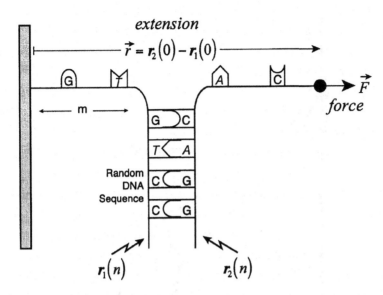

Figure 1. Schematic of DNA unzipping experiments. The sugar-phosphate backbones, with complementary base pairing, are represented by the functions $r_1(n)$ and $r_2(n)$, where n indexes base pairs. Experiments can be carried out either at constant extension $r_1(0) - r_2(0)$ or at constant force $\vec{F}$.

the extension. Precise analytic predications are indeed possible in the vicinity of F_c, even when sequence disorder is present [9, 14]. Third, experiments which use force as a control parameter can probe interesting behavior of the unzipping fork for $F < F_c(T)$ (analogous to "wetting" phenomena in bulk phase transitions [16]) and anomalous fork dynamics for $F > F_c(T)$. Because of significantly different bonding strengths of the nucleotide pairs A:T (two hydrogen bonds) and G:C (three hydrogen bonds), heterogeneity dominates unzipping over a large region of the phase diagram, for $\frac{1}{2}F_c \lesssim F \lesssim \frac{3}{2}F_c$, inside the region bounded by the dashed lines in Fig. 2.

2. Thermal Denaturation of DNA

In this Section we discuss the thermal denaturation of DNA at zero force (vertical arrow in Fig. 2). This transition plays an important role in the polymerase chain reaction (PCR), where it is applied cyclically to amplify minute amounts of DNA. The basic physical ideas driving thermal denaturation (worked out in the 1960s) are described in Refs. [17–19]. The simplest models invoke the Ising-like description of DNA

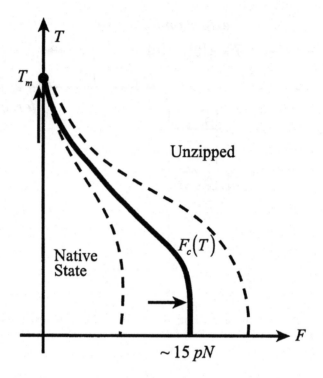

Figure 2. Phase diagram for DNA denaturation in the force-temperature plane. Conventional thermal denaturation occurs at zero force as T approaches the melting temperature T_m. Force-induced denaturation, or unzipping occurs on the path indicated by the horizontal arrow. Sequence heterogeneity dominates the physics between the two dashed crossover lines.

shown in Fig. 3a, where right-facing spins represent bonded nucleotides in a closed "helix" state and left-facing spins the open "coil" state. The magnetic field of the equivalent Ising model represents a (temperature-dependent) nucleotide bonding free energy and the Ising exchange energy determines an "initiation factor" at the boundaries between the helix and coil regions. The statistical mechanics of this one-dimensional Ising model can be described entirely in terms of kink excitations associated with transitions between domains of up and down spins, which here represent helix/coil boundaries. Rapid variations in DNA from a helix-dominated to a coil-dominated state are predicted by such models [19], corresponding to the magnetic field changing sign in the equivalent Ising system. However, it is well-known that the 1d Ising model does not have a real phase transition at finite temperature, basically because the

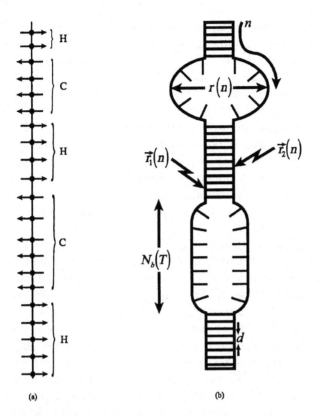

Figure 3. Models of thermal denaturation. (a) Ising model, where right-pointing spins represent hydrogen-bonded "helix" segments, and left-pointing spins represent denatured "coil" segments. (b) More sophisticated polymer model of denaturation whose statistical mechanics is mapped onto a problem in three-dimensional quantum mechanics. $N_b(T)$ is the number of base pairs associated with a typical denatured "coil" segment.

finite energy cost of kinks at zero external field is overwhelmed by the entropy gained by creating them [20]. An improved theory results when the three-dimensional entropy of wandering loops in the coil sections (see Fig. 3b) is taken into account. This can be done by approximating the loops as ideal random walks [17], or improving this estimate by taking self-avoidance of the loops into account [21]. The effect on the statistical mechanics is to produce a long range interaction which binds kinks together and leads to a continuous finite temperature phase transition [17–19]. Recent work by Kafri *et al.* incorporates additional self-avoidance between coil sections and the rest of the DNA chain (important for large molecules) and argues that the thermal denaturation transition then becomes first order [22].

Heterogeneity of the bonding between base pairs (G:C bonds are stronger than A:T ones) is neglected in the treatments above. Simple helix-coil descriptions with heterogeneity are governed by the statistical mechanics of the 1d random field Ising model. A reasonable guess for more realistic models with heterogeneity is that while the transition may be less sharp (weakly bound regions melt before more tightly bound ones), a real finite temperature phase transition survives. Lyubchenko *et al.* [23] have calculated numerically the effect of sequence heterogeneity on the DNA differential melting curves for the 5375 nucleotides of the virus $\phi X 174$. For a recent theory of thermal denaturation of heterogeneous DNA, see Ref. [24].

In the remainder of this section, we discuss how thermal denaturation of "homogeneous" DNA with identical base pairs can *also* be understood by mapping the statistical mechanics onto the delocalization of a quantum mechanical particle in three dimensions. This is the approach we will generalize in Sec. 3 to study force-induced naturation.

Referring to Fig. 3b, we parameterize the positions of the two (antiparallel) sugar-phosphate backbones of DNA in three dimensions by the functions $r_1(n)$ and $r_2(n)$, where n is an integer indexing N base pairs ($0 \leq n \leq N$). We model each strand separately as a Gaussian random coil [25] . For example, considering strand 1 in isolation, the probability of a particular polymer conformation $r_1(n)$ is proportional to $e^{-F_1[r_1(n)]/T}$ (we use units such that $k_B = 1$), where

$$F_1[r_1(n)] = \frac{1}{2}K \int_0^{Nd} \left(\frac{dr_1(s)}{ds}\right)^2 \, ds. \tag{1}$$

Here we have used a convenient continuum notation as shorthand for a discrete sum over n and replaced n by the arclength $s = nd$, where d is the spacing between nucleotide monomers. To relate the spring constant K to the single strand persistence length, use (1) to evaluate $\langle |\vec{r}_1(Nd) - \vec{r}_1(0)|^2 \rangle$, with the result

$$\langle |\vec{r}_1(Nd) - \vec{r}_1(0)|^2 \rangle = \frac{3TNd}{K}. \tag{2}$$

From the definition of persistance length ℓ, namely $\langle |\vec{r}_1(Nd) - \vec{r}_1(0)|^2 \rangle \equiv \ell Nd$, we have [26]

$$K = \frac{3T}{\ell}, \tag{3}$$

where $\ell \approx 10$Å for single-stranded DNA at physiological temperatures. For a more microscopic "worm-like chain" model with bending rigidity κ (assuming a small Debye screening length), we have [27, 28]

$$\ell \approx \frac{2\kappa}{T}. \tag{4}$$

Here, and in what follows, we use boldface vectors $\mathbf{r}(s)$ to denote entire polymer configurations, and conventional vectors such as $\vec{r}(Nd)$ to denote the position of a polymer endpoint.

To describe a DNA duplex, we bring the two strands together, neglect torsional rigidity and the helical nature of the bonding [28], and write the total free energy as

$$F[\mathbf{r}_1(s), \mathbf{r}_2(s)] = \frac{K}{2} \int_0^{Nd} \left[\left(\frac{d\mathbf{r}_1(s)}{ds} \right)^2 + \left(\frac{d\mathbf{r}_2(s)}{ds} \right)^2 \right] ds$$
$$+ \int_0^{Nd} U[\mathbf{r}_1(s) - \mathbf{r}_2(s)] \ ds, \tag{5}$$

where $U[\mathbf{r}_1(s) - \mathbf{r}_2(s)]$ is a potential which binds the nucleotides in different strands together. We assume that the complimentarity of the nucleotide sequences is sufficient to bring the strands into registry, but will otherwise neglect the effect of sequence disorder in the remainder of this section. If we now pass to sum and difference variables.

$$\mathbf{R}(s) = \frac{1}{2}[\mathbf{r}_1(s) + \mathbf{r}_2(s)], \tag{6}$$

$$\mathbf{r}(s) = \mathbf{r}_2(s) - \mathbf{r}_1(s), \tag{7}$$

the coarse-grained free energy decouples,

$$F[\mathbf{R}(s), \mathbf{r}(s)] = K \int_0^{Nd} ds \left(\frac{d\mathbf{R}}{ds} \right)^2 + \frac{1}{2} g \int_0^{Nd} ds \left(\frac{d\mathbf{r}}{ds} \right)^2$$
$$+ \int_0^{Nd} ds \ U[\mathbf{r}(s)], \tag{8}$$

where $g = K/2$.

The partition function associated with (8) is a path integral of $e^{-F(\mathbf{R},\mathbf{r})/T}$ over the functions $\mathbf{R}(s)$ and $\mathbf{r}(s)$. The part associated with $\mathbf{R}(s)$ is simply the partition sum of an unconstrained Gaussian coil. The remaining functional integral over $\mathbf{r}(s)$ contains a denaturation transition. To pursue this point, we assume for simplicity that $\mathbf{r}(s = 0) = \vec{0}$ and

$r(s = Nd) = \vec{r}$ (closed boundary conditions at one end, open at the other), and evaluate

$$Z(\vec{0}, \vec{r}; N) = \int_{r(0)=\vec{0}}^{r(Nd)=\vec{r}} \mathcal{D}r \exp\left[-\frac{g}{2T} \int_0^{Nd} \left(\frac{dr}{ds}\right)^2 ds - \frac{1}{T} \int_0^{Nd} U[r(s)] \right] ds. \tag{9}$$

As in polymer adsorption problems, it is helpful to view Eq. (9) as a Feynman path integral for a quantum mechanical particle in imaginary time [27]. Indeed, this partition function can be rewritten as a quantum mechanical matrix element [29],

$$Z(\vec{0}, \vec{r}; N) = \langle \vec{r} | e^{-\hat{H}Nd/T} | \vec{0} \rangle, \tag{10}$$

where the effective quantum Hamiltonian is

$$\hat{H} = \frac{-T^2}{2g} \nabla^2 + U(\vec{r}) \tag{11}$$

and $|0\rangle$ and $\langle \vec{r} |$ are respectively *ket* and *bra* vectors localized at $\vec{0}$ and $\vec{r}$. Note that temperature plays the role of $\hbar$ and $g = \frac{1}{2}K = \frac{3T}{2\ell}$ represents the mass of a fictitious quantum mechanical particle in a potential $U(\vec{r})$. In the language of statistical mechanics, we have reduced this one-dimensional problem to diagonalizing a transfer matrix given by $\hat{T} = e^{-\hat{H}d/T}$. We shall focus on the particularly simple binding potential illustrated in Fig. 4, namely

$$U(\vec{r}) = \begin{cases} \infty, & r < c \\ -U_0, & c < r < b \\ 0, & b < r, \end{cases}$$

with $U_0 = V_0/d$, where V_0 is the average bonding energy per nucleotide. Upon inserting a complete set of energy eigenfunctions $|n\rangle = \phi_n(\vec{r})$ with eigenvalues ϵ_n into Eq. (10), we see that this conditional partition function may be rewritten as

$$Z(\vec{0}, \vec{r}; N) = \sum_n \langle \vec{r} | e^{-\hat{H}Nd/T} | n \rangle \langle n | 0 \rangle = \sum_n \phi_n^*(\vec{0})\phi_n(\vec{r}) e^{-\epsilon_n Nd/T}. \tag{12}$$

The ground state dominates in the limit $N \to \infty$, and we have

$$Z(\vec{0}, \vec{r}; N) \approx \phi_0(\vec{0})\phi_0(\vec{r}) e^{-\epsilon_0 Nd/T}, \tag{13}$$

where the (nodeless) ground state eigenfunction can be chosen to be real.

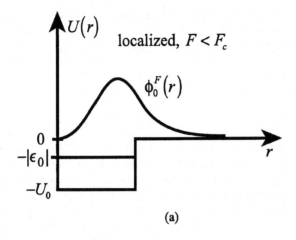

(a)

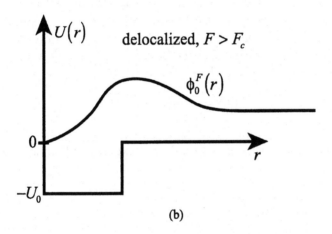

(b)

Figure 4. Potential well and ground state eigenfunctions for forced-induced denaturation when (a) $F < F_c$ and (b) $F > F_c$. These (right) eigenfunctions give the
probability of a separation $\vec{r}$ for the unzipped ends of the DNA. For $F < F_c$, the
ground state eigenfunction decays exponentially to zero outside the well, which models the hydrogen bonds and stacking energies which hold base pairs together. For
$F > F_c$, the eigenfunction tends to a nonzero constant for large r, and the two DNA
strands fall apart.

The quantum mechanics of simple square well potentials like Eq. (12)
in three dimensions is well understood [30]. We focus for simplicity on

the case $c << b$ (small hard core diameter) but the same qualitative discussion applies more generally. The key parameter controlling the eigenfunctions and eigenvalues is the dimensionless ratio

$$Q = \frac{2gU_0 b^2}{T^2} \qquad (14)$$

of the well depth U_0 to the "zero point energy" $\frac{T^2}{2gb^2}$. This quantum zero point energy represents the lost entropy associated with a confined DNA duplex. From the perspective of quantum mechanics, the loop excitations shown in Fig. 3b represent transient excursions (in imaginary time) out of the well. When $Q >> 1$ (low temperatures), the binding energy dominates the thermal fluctuations favoring denaturation and there are many bound (i.e., localized) eigenstates of $\hat{H}$ beneath a continuum of extended eigenstates. As Q decreases, more and more states delocalize until only the single localized ground state shown in Fig. (4a) is present. Eventually, at temperatures high enough so that $Q \leq Q^*$, even the ground state delocalizes [31]. For the simple problem discussed here $Q^* = \frac{\pi^2}{4}$ (i.e., $T_m = 8U_0 b^2/\pi^2 g$) [30]. However, a similar Q^* of order unity is expected for a broad class of similar potentials. Near the transition temperature T_m, the ground state energy approaches the continuum quadratically with temperature [27, 30]

$$\epsilon_0(T) \approx -\text{const.}|T_m - T|^2, \qquad (15)$$

where the melting temperature T_m is defined by $Q(T_m) = Q^*$. For large r, the ground state wave function decays exponentially, $\phi_0(\vec{r}) \sim e^{-\kappa_0|\vec{r}|}$, where

$$\kappa_0^{-1}(T) = \sqrt{\frac{-T^2}{2g\epsilon_0(T)}} \sim \frac{1}{(T_m - T)} \qquad (16)$$

gives the spatial extent (in three dimensions) of the bubbles shown schematically in Fig. 3b. The *number* of base pairs N_b in a bubble (see Fig. 3b) can be estimated by identifying N_b with the value of N needed to achieve ground state dominance in Eq. (12): We expect that boundary effects become unimportant when the number of base pairs exceeds the number in a typical bubble. The ground state will dominate once the number of nucleotides is larger than several bubble sizes. Since the energy of the first excited state for $Q \gtrsim Q^*$ is zero, we have

$$Z(\vec{0}, \vec{r}; N) \approx \phi_0(\vec{0})\phi_0(\vec{r})e^{\frac{|\epsilon_0|Nd}{T}} \left[1 + \mathcal{O}\left(e^{\frac{-|\epsilon_0|Nd}{T}}\right)\right], \qquad (17)$$

from which it follows that

$$N_b \approx \frac{T}{d|\epsilon_0(T)|} \sim \frac{1}{(T_m - T)^2}. \tag{18}$$

We should stress that the above results are obtained in an approximation which neglects self-avoidance effects important for large molecules close to T_m. We expect, however, qualitatively similar behavior even when self-avoidance is taken into account. We now use similar methods to treat force-induced denaturation in Section 3.

3. Force-induced Denaturation

A force $\vec{F}$ applied across the two strand endpoints (see Fig. 1) adds an energy $-\vec{F} \cdot \vec{r}$, and Eq. (9) becomes

$$Z(\vec{0}, \vec{r}; N) = \int_{\mathbf{r}(0)=\vec{0}}^{\mathbf{r}(Nd)=\vec{r}} D\mathbf{r}(s) \exp\left[-\frac{g}{2T} \int_0^{Nd} \left(\frac{d\mathbf{r}}{ds}\right)^2 ds - \frac{1}{T} \int_0^{Nd} U_s[\mathbf{r}(s)] ds \right. $$
$$\left. + \frac{\vec{F}}{T} \cdot \int_0^{Nd} \frac{d\mathbf{r}}{ds} ds \right]. \tag{19}$$

We assume the attached strands of DNA can swivel freely to relax twist on an experimental time scale, so that we can neglect the helical nature of the duplex state when the force is applied. In addition to the force term (note that the lower limit does not contribute with our boundary conditions), we have added a subscript s to the binding potential, $U[\mathbf{r}(s)] \to U_s[\mathbf{r}(s)]$, to emphasize that the depth and size of potentials like that in Fig. 4 will depend on the particular base pair for heterogeneous DNA. After taking over results for a virtually identical problem [32] of flux line depinning by a columnar pin in a transverse magnetic field in high T_c superconductors, we obtain [9, 10]

$$T \frac{\partial Z(\vec{0}, \vec{r}; N)}{\partial Nd} = \frac{-T^2}{2g} \left(\vec{\nabla} - \frac{\vec{F}}{T} \right)^2 Z(\vec{0}, \vec{r}; N) + U_{Nd}(\vec{r}) Z(\vec{0}, \vec{r}; N), \tag{20}$$

which can be integrated (subject to the boundary condition $Z(\vec{0}, \vec{r}; 0) = 1$) to give a result with the form of Eq. (10), where

$$\hat{H}(\vec{F}) = \frac{-T^2}{2g} \left(\vec{\nabla} - \frac{\vec{F}}{T} \right)^2 + U_{Nd}(\vec{r}). \tag{21}$$

Equation (20) with an s-dependent potential $U_s[\vec{r}]$ can be studied (via a mapping onto a Burgers equation) by the methods of Ref. [33], where it arose in a study of vortex depinning from *fragmented* columnar pins. In this section, we neglect the s-dependence (letting $U_{Nd}(\vec{r}) \to U(\vec{r})$) and illustrate force-induced denaturation for the homopolymer duplex discussed in Sec. 2. Because $\vec{F}$ appears as a constant *imaginary* "vector potential," Eq. (21) represents a non-Hermitian generalization of Schroedinger's equation. To establish the existence of a first-order transition, we write the solutions of (20) using the right and left eigenfunctions $\{\phi_{nR}^F(\vec{r})\}$ and $\{\phi_{nL}^F(\vec{r})\}$ (with a set of common eigenvalues $\{E_n(F)\}$) of $\hat{H}(\vec{F})$,

$$Z(\vec{0}, \vec{r}; N) = \sum_n \phi_{nL}^F(\vec{0}) \phi_{nR}^F(\vec{r}) e^{-E_n(F)Nd/T}. \tag{22}$$

We again invoke ground-state dominance and focus on the behavior of the lowest eigenvalue, which determines the duplex free energy per unit length $a(F)$ via

$$a(F) = -T \lim_{N \to \infty} \frac{1}{Nd} \ln[Z(N, F)], \tag{23}$$

where $Z(N, F)$ is defined by integrating over $\vec{r}$,

$$Z(N, F) = \int d^3r \, Z(\vec{0}, \vec{r}; N), \tag{24}$$

since we work at constant force instead of constant extension.

For F less than a critical value F_c (determined below), we can obtain the (right) eigenfunctions $\phi_{nR}^{\vec{F}}(\vec{r})$ for $\vec{F} \neq 0$ from localized eigenfunctions $\phi_n(\vec{r})$ for the Hermitian case $\vec{F} = 0$ via a "gauge transformation" [32]. Indeed, it is easy to check that

$$\phi_{nR}^{\vec{F}}(\vec{r}) = e^{\frac{\vec{F} \cdot \vec{r}}{T}} \phi_n(\vec{r}) \tag{25}$$

solves

$$\hat{H}(\vec{F}) \phi_{nR}^{\vec{F}}(\vec{r}) = E_n \phi_n^F(\vec{r}) \tag{26}$$

(provided $\phi_n(\vec{r})$ satisfies $\hat{H}(\vec{0}) \phi_n(\vec{r}) = \epsilon_n \phi_n(\vec{r})$) with the *same* eigenvalue, $E_n = \epsilon_n$. In particular, we obtain from Eq. (23) a negative $\vec{F}$-independent free energy per unit length $a = \epsilon_0 = -|\epsilon_0|$ for small $\vec{F}$. If, however, a localized ground-state eigenfunction for the Hermitian $\vec{F} = 0$ problem decays like $\phi_0(\vec{r}) \sim e^{-\kappa_0|\vec{r}|}$ for large r, the new eigenfunctions (25) are only normalizable provided $F \leq F_c(T) = T\kappa_0(T)$. Note that $\phi_{0R}^{\vec{F}}(\vec{r})$, which gives the probability of finding an extension $\vec{r}$, [32] is displaced in the direction of $\vec{F}$, as indicated schematically in Fig. 4a. To determine what happens for $F > F_c$, we check for a nodeless *extended* ground-state wavefunction, as indicated in Fig. 4b. If $\phi_{0R}^F(\vec{r}) \to$ const. > 0 for large r (i.e., the DNA duplex falls apart), and $\lim_{r \to \infty} U(\vec{r}) = 0$, evaluating Eq. (22) in this limit gives $\hat{H}(F)\phi_{0R}^F(\vec{r}) = \frac{-F^2}{2g}\phi_{0R}^F(\vec{r})$, so that $a(F) = E_0(F) = \frac{-F^2}{2g}$. The two free energies $a(F)$ for $F < F_c$ and $F > F_c$ are plotted in Fig. 5 [32]. At F_c the two curves intersect at a nonzero angle, the classic signature of a first-order phase transition. The latent heat of this transition might be observable in a solution of many DNA duplexes pulled apart by, say, beads of opposite charge in an electric field. Since the first-order transition curve $F_c(T)$ is given by $-F_c^2/2g = -|\epsilon_0(T)|$, where ϵ_0 is the smallest eigenvalue of the transfer matrix for $F = 0$, we find using Eq. (15) that

$$F_c(T) \propto \text{const.}|T_m - T| \qquad (27)$$

near the melting transition for this model.

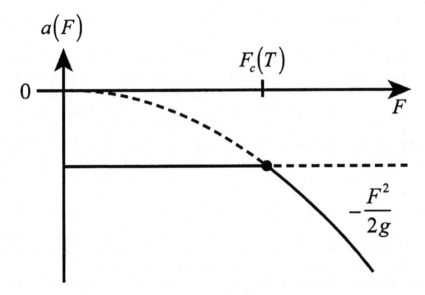

Figure 5. The constant free energy $a = -|\epsilon_0|$ per base of duplex DNA compared to the parabolic free energy $a(F) = -F^2/2g$ appropriate for two separated strands. See the text for a discussion of more sophisticated freely-jointed chain and worm-like chain models of this free energy. The two free energies cross with discontinuous slope at $F_c(T)$, the location of the first-order unzipping transition.

Even though force-induced denaturation is a first-order transition, there are diverging precursors in the form of DNA unzipping (see Fig. 1). The probability of an endpoint separation $|\vec{r}|$ in the limit $N \to \infty$ is given integrating $|\vec{r}|$ weighted by the ground state eigenfunction,

$$\phi_{0R}^{\vec{F}} = e^{\frac{\vec{F} \cdot \vec{r}}{T}} \phi_0(\vec{r})$$

$$\underset{r \to \infty}{\simeq} e^{\frac{\vec{F} \cdot \vec{r}}{T}} e^{-\kappa_0 |r|}. \tag{28}$$

It is then straightforward to show that $\langle |\vec{r}| \rangle$ diverges as $F \to F_c^-(T)$ [32]

$$\langle |\vec{r}| \rangle \sim \frac{1}{F_c(T) - F}. \tag{29}$$

Upon examining corrections to ground state dominance as in the analysis of Eq. (19), we find that the average number m of unzipped monomers diverges, as shown in Fig. 6a,

$$\langle m \rangle \simeq \frac{T}{d(|\epsilon_0| - F^2/2g)}$$

$$\sim \frac{1}{F_c(T) - F}. \tag{30}$$

The first-order force-induced DNA denaturation transition can be understood more generally, independent of the mapping onto quantum mechanics. If $a_1(T, F)$ is the free energy per unit length of one of the single-stranded DNA "handles" shown in Fig. 7, and $a_0(T)$ is the (force-independent) free energy per unit length of the double-stranded DNA which coexists in a macroscopically unzipped state, then the condition for two-phase coexistence across the "meniscus" or unzipping fork at the critical force F_c is [14]

$$a_0(T) = 2a_1(F_c, T). \tag{31}$$

The physics of the handle free energy $a_1(F, T)$ is relatively simple, provided the force is large enough to neglect self-hybridization and self-avoidance of the single-stranded DNA. For example, we could take $a_1(T, F)$ to be the free energy of a freely-jointed chain with persistance length ℓ [25]

$$a_1(T, F) \propto -\frac{T}{\ell} \ln\left(\frac{T \sinh(F\ell/T)}{F\ell}\right), \tag{32}$$

which is proportional to $-\ell F^2/T$ for small F. Alternatively one could use a more sophisticated "worm-line chain" approximation valid over a

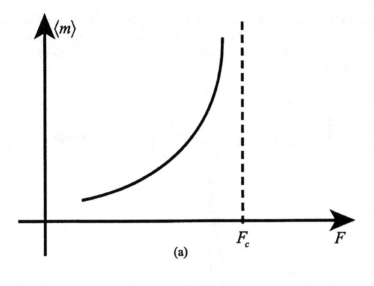

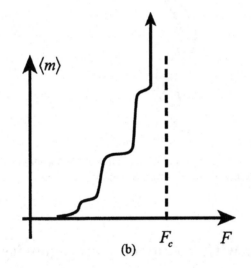

Figure 6. Thermal average $\langle m \rangle$ of the location of the unzipping fork. (a) Divergence of $\langle m \rangle$ at F_c for a homopolymer. (b) Jumps and plateaus in $\langle m \rangle$ (with, on average, a much strong divergence) in the presence of sequence disorder.

larger range of forces [34]. With an acceptable approximation to $a_1(T, F)$ in hand (if necessary, this function could be obtained directly from integrating single strand force-extension curves [35]), one can use the experimentally determined phase boundary $F_c(T)$ for unzipping in conjunction with Eq. (31) to explore the temperature dependence of the free energy $a_0(T)$ of double-stranded DNA.

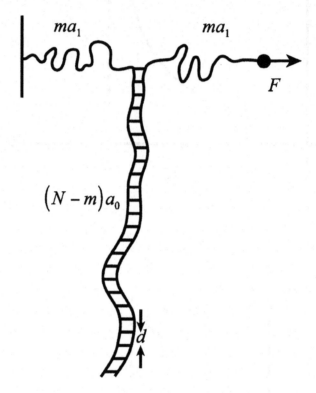

Figure 7. Two-phase coexistence of duplex DNA with free energy oer base pair a_0 and two unzipped "handles," each with free energy per base a_1.

4. Unzipping with Sequence Heterogeneity

The effect of sequence heterogeneity on thermal denaturation of DNA at zero force is a subtle and still not completely resolved problem, [19, 23, 24]. Understanding inhomogeneous base pairing energies near the first-order force-induced unzipping transition appears to be more tractable. The integrated effect of sequence randomness on the energy landscape for long strands of DNA does not change the underlying first-order transition, because it only produces corrections of order $\frac{1}{\sqrt{N}}$ to the free energies per unit length discussed above, where N is the total number of base pairs. However, sequence randomness has drastic consequences for quantities like $\langle m \rangle$ which characterize the unzipping process itself [9, 14]. Here, we simply sketch the overall picture. Detailed calculations and justifications for approximations (such as the neglect of bubbles in DNA) can be found in Ref. [14].

Consider the energy $\epsilon(m)$ associated with the conditional partition sum of a DNA duplex which has been unzipped by m base pairs. If the DNA is a homopolymer, consisting exclusively of A:T or G:C base pairs, the total energy can be read off from Fig. 7,

$$
\begin{aligned}
\mathcal{E}(m) &= 2ma_1 + (N - m)a_0 \\
&= a_0N + fm
\end{aligned}
\tag{33}
$$

where

$$
f \equiv 2a_1(F) - a_0 \tag{34}
$$

can be expanded to give $f \approx c\frac{2da_1}{dF}|_{F_c}(F_c - F)$ near the unzipping transition. If we now add sequence randomness, a reasonable model for the energy of unzipping is

$$
\begin{aligned}
\Delta\mathcal{E}(m) &\equiv \mathcal{E}(m) - Na_0 \\
&= fm + \int_0^m dm'\eta(m'),
\end{aligned}
\tag{35}
$$

where deviations from the average A:T or G:C content are represented by the function $\eta(m')$ in a convenient continuum notation. To the extent that most DNA sequences which code for proteins have deviations from the average G:C/A:T content describable as a random walk [36], we expect that the statistical properties of $\eta(m)$ at large distances are those of a white-noise random variable,

$$
\overline{\eta(m)\eta(m')} = \Delta_0\delta(m - m') \tag{36}
$$

where the overbar represents a "disorder average" over all possible sequences of base pairs. Because G:C and A:T pairing energies differ by an amount of order T, $\Delta_0 \simeq T^2$, where we have used the temperature $T \simeq$ 293–310°K characteristic of most experiments to characterize this quantity. The partition function associated with this simple one-dimensional model involves integrating over all possible unzipping lengths m,

$$
\mathcal{Z}(f) = \int_0^\infty dm e^{-\Delta\mathcal{E}(m)/T}. \tag{37}
$$

The unzipping energy in Eq. (35) is plotted for a particular base pair sequence (with $\Delta_0 = T^2$) in Fig. 8, with $f/T = 0.01$. The straight line is the energy of a *homopolymer* duplex with $\Delta_0 = 0$ and

$$\Delta\mathcal{E}(m) = fm. \tag{38}$$

In this case we find immediately from Eq. (37) that the thermally averaged degree of unzipping is

$$\langle m \rangle = -T\frac{d \ln \mathcal{Z}(f)}{df}$$
$$= \frac{T}{f} \sim \frac{1}{(F_c - F)}, \tag{39}$$

in agreement with Eq. (30). There are large fluctuations about this average value, as one can check by evaluating

$$\sqrt{\langle (m - \langle m \rangle)^2 \rangle} = T/f = \langle m \rangle. \tag{40}$$

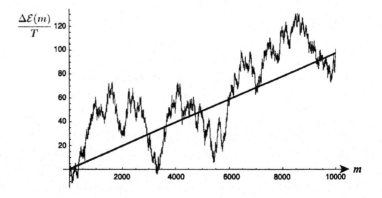

Figure 8. Energy landscape for $F \lesssim F_c$ for a homopolymer duplex (straight line) and with the addition of inhomogeneous base pairing energies due to a particular DNA sequence (jagged line). Here we take $\Delta_0 = T^2$ and $f/T = 0.01$.

The rugged landscape in Fig. 8 for $\Delta_0 \neq 0$ is more interesting. Although the positive background slope determined by f insures that the DNA remains macroscopically unzipped, the integrated effect of sequence randomness produces deviations from a straight line which scale like $\sqrt{m}$. These fluctuations lead to deep minima at nonzero m with large energy barriers in between. A downward fluctuation in this "integrated random walk" energy landscape at position m corresponds to an energy of order

$$\Delta\epsilon(m) \simeq fm - \sqrt{\Delta_0 m}. \tag{41}$$

Upon minimizing over m, we obtain the estimate

$$\overline{\langle m \rangle} \sim \frac{\Delta_0}{f^2} \sim \frac{1}{(F_c - F)^2} \tag{42}$$

a result confirmed by more elaborate calculations [9, 14]. Here, the overbar represents an average over a quenched random distribution of DNA sequences. It can also be shown that thermal fluctuations about this average are more constrained than for homopolymers, in the sense that

$$\sqrt{\overline{\langle (m - \langle m \rangle)^2 \rangle}} \sim \frac{1}{f^{3/2}} \sim \overline{\langle m \rangle}^{3/4}. \tag{43}$$

Thus, in contrast to homopolymers, where $\langle (m - \langle m \rangle)^2 \rangle / \langle m \rangle^2 = \mathcal{O}(1)$ we now have

$$\overline{\langle (m - \langle m \rangle)^2 \rangle} / (\overline{\langle m \rangle})^2 = \mathcal{O}(1/\overline{\langle m \rangle}^{1/2}). \tag{44}$$

A typical energy barrier associated with disorder-dominated unzipping is $f \langle m \rangle + \sqrt{\Delta_0 \langle m \rangle} \sim \frac{\Delta_0}{f}$. Results such as (42) apply when this scale exceeds T, i.e., for f less than a crossover force f_x,

$$f < f_x = \frac{\Delta_0}{T}. \tag{45}$$

Because $\Delta_0 = \mathcal{O}(T^2)$ is large for G:C and A:T base pairs, the range of forces where sequence heterogeneity dominates the physics is the large region bounded by the left-hand dashed line in Fig. 2. Note that disordered dominated prediction for $\overline{\langle m \rangle}$ in Eq. (42) diverges with a power which is *twice* as large as the result (39) appropriate for thermally equilibrated homopolymers.

Results such as Eq. (42) apply only to a quenched average over an entire library of different DNA sequences unzipped in parallel. We can understand the behavior for $F \lesssim F_c(T)$ for the more experimentally accessible case of a *particular* squence from Fig. 9, which shows the energy landscape for two identical sequences with the biases, $f/T = 0.01$ and $f/T = 0 = 0.006$. The degree of unzipping $\langle m \rangle$ is dominated by the minimum indicated by an arrow in each case. Somewhere between $f = 0.01$ and $f = 0.006$, the average of $\langle m \rangle$ jumps from one minimum to the next. More detailed calculations [14] reveal an entire sequence of

84

jumps and plateaus, as illustrated in Fig. 6b. Although a best fit to a power law of this irregular curve should reveal the exponent of Eq. (42), the plateaus and jumps themselves represent a rough "fingerprint" of the individual sequence. See Ref. [14], where the statistical distribution of plateaus and jumps are evaluated using the techniques of LeDoussal *et al.* [37].

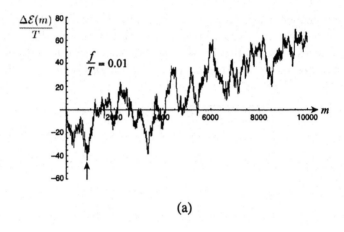

(a)

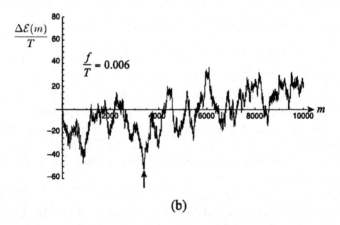

(b)

Figure 9. Energy landscapes two identical DNA sequences with $\Delta_0 = T^2$ and (a) $f/T = 0.01$ and (b) $f/T = 0.006$. The arrow indicates the energy minimum which dominates the value of $\langle m \rangle$ in each case.

5. Dynamics of Unzipping

What happens for $F \gtrsim F_c(T)$? In Figure 10 we show an energy landscape (for the same sequence as in Fig. 8) with $f/T = -0.01$, together with the purely downhill landscape of the corresponding homopolymer. Because the average free energy per base pair of the DNA duplex exceeds the combined free energies per base of the two single-stranded DNA "handles" in Fig. 7, the system is unstable to complete unzipping. It is then appropriate to discuss the *dynamical* process by which unzipping proceeds on this downhill path.

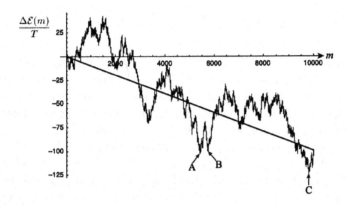

Figure 10. Downhill energy landscape appropriate for $F \gtrsim F_c(T)$. Although the overall slope is negative, there are many traps, such as those at positions A, B, and C. Energy barriers between these traps scale as $\sqrt{\Delta_0 m}$, where m is the number of bases between traps.

Subject to a number of simplifying conditions discussed in Ref. [14], the time dependence of the unzipping fork can be described by an overdamped Langevin equation,

$$\frac{dm(t)}{dt} = -\Gamma \frac{d\mathcal{E}(m)}{dm} + \zeta(t), \tag{46}$$

where Γ is a damping constant ($\Gamma = 1/(\tau T)$, where τ is a microscopic relaxation time) and the noise correlations are

$$\langle \zeta(t)\zeta(t') \rangle = 2T\Gamma\delta(t - t'). \tag{47}$$

Upon substituting for the energy landscape $\mathcal{E}(m)$ from Eq. (35), we consider the case $f < 0$ and obtain

$$\frac{dm(t)}{dt} = \Gamma|f| - \Gamma\eta(m) + \zeta(t), \tag{48}$$

which is the equation of motion for a particle with coordinate $m(t)$ executing one-dimensional biased diffusion in a random force field proportional to $\eta(m)$.

A great deal is known about this problem [38–40]. In the absence of sequence heterogeneity ($\Delta_0 = 0$ in Eq. (36)), $m(t)$ exhibits diffusion with drift at long times, i.e.,

$$\lim_{t \to \infty} \frac{\langle m(t) \rangle}{t} = v \tag{49}$$

$$\lim_{t \to \infty} \frac{\langle [m(t) - \langle m(0) \rangle]^2 \rangle}{t} = 2D \tag{50}$$

with a well-defined drift velocity $v = \Gamma |f|$ and diffusion constant $D = \Gamma T$. This is the expected behavior for the straight line homopolymer energy landscape shown in Fig. 10. Sequence heterogeneity, however, has a dramatic effect on the dynamics unless $|f|$ is large. Indeed, the integrated random walk landscape in Fig. 10 has many deep minima, even though the average slope is negative. When $|f|$ is small, typical energy barriers to travel a distance m scale like $\sqrt{\Delta_0 m}$, as is reflected in the small barrier connecting the minima labelled A and B and the much larger barrier between these minima and the more distant minimum C. A detailed analysis [38–40] reveals three types of anomalous dynamics, depending on the dimensionless parameter

$$\mu = \frac{2T|f|}{\Delta_0}. \tag{51}$$

Right at the unzipping transition, $\mu = |f| = 0$ and the unzipping fork wanders sub-diffusively according to [38]

$$\overline{\langle |m(t) - m(0)|^2 \rangle} \underset{t \to \infty}{\simeq} \frac{T^4}{\Delta_0^2} \ln^4(t/\tau), \tag{52}$$

because of trapping effects arising from the large energy barriers. The conventional diffusion constant D, as defined by Eq. (50), vanishes. When $0 < \mu < 1$, the unzipping fork drifts downhill, but does so sublinearly with time,

$$\overline{\langle m(t) \rangle} \underset{t \to \infty}{\simeq} \text{const. } t^\mu, \tag{53}$$

so the usual drift velocity v defined by (49) vanishes, with an additional anomaly in the spread about this average drift [39, 40]. When this bias is

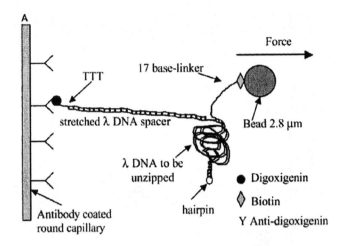

Figure 11. Experimental setup for the force DNA unzipping experiments of Danilowicz *et al.* [42]. The unzipping trajectories of dozens of genetically identical lambda phage DNAs attached to a thin capillary tube are tracked. A constant force is generated by attaching magnetic beads and placing the apparatus in a magnetic field gradient.

large enough so that $1 < \mu < 2$, one recovers a well-defined drift velocity v, but with superdiffusive spreading,

$$\overline{\langle[m(t) - \langle m(t)\rangle]^2\rangle} \underset{t\to\infty}{\sim} t^{2/\mu}. \tag{54}$$

Conventional diffusion with drift is only recovered for forces large enough so that $\mu > 2$, i.e., for

$$-f > f'_x = \frac{\Delta_0}{T}. \tag{55}$$

This critical force is the same order of magnitude as the crossover scale defined *below* the unzipping transition by Eq. (45) and leads to the right-hand dashed line in Fig. 2. The parameters of DNA are such that anomalous unzipping appears in a very large range (compared to typical disorder effects in conventional critical phenoman [41]), roughly $\frac{1}{2}F_c \lesssim F \leq \frac{3}{2}F_c$.

6. Pauses, Jumps and Concluding Remarks

Experiments which study the mechanical denaturation of lambda phage DNA at constant force have recently been carried out in the group of

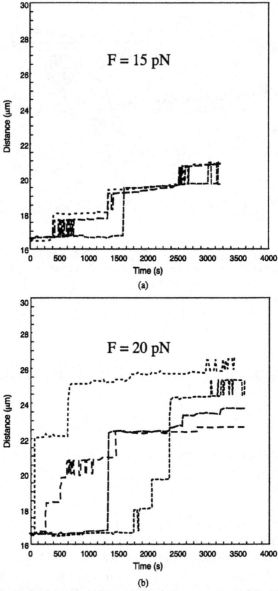

Figure 12. Unzipping histories obtained from the apparatus of Fig. 11 for (a) three beads with $F = 15$ *pN* and (b) four beads with $F = 20$ *pN*. Note the long-time scale (of order 1 hour) and the many jumps and plateaus. The oscillations arise when the unzipping fork encounters two nearly degenerate energy minima. The same plateaus recur for different beads at the same force, suggesting that these are due to sequence disorder (from Ref. [42]).

Mara Prentiss [42]. Digoxigenin linkers attach several dozen identical copies of this DNA along a long thin capillary tube. Each DNA to be unzipped is offset by an additional DNA spacer (see Fig. 11). Biotin links the other end of the DNA to magnetic beads. Unzipping at constant force is achieved by applying a uniform magnetic field gradient, causing the beads to stretch out at right angles to the capillary tube, like a series of flags on a flagpole.

Figure 12a records some unzipping histories at $T = 20°C$ for three different beads at different heights on this "flagpole" with $F = 15\ pN$, corresponding approximately to the critical unzipping force for the front half of the sequence for phage lambda. (The back half of this DNA has a lower G:C content.) The beads are well separated, and are clearly subject to microscopically different thermal histories. Nevertheless, there are distinctive pauses (some lasting 15 minutes or more) and jumps common to the different beads. As the individual DNA's unzip, the beads occasionally jump back and forth, suggesting two nearly degenerate minima in the energy landscape about 1000 base pairs apart. This behavior is similar to that found for much shorter RNA hairpains in Ref. [5]. The long pauses before unzipping resumes are consistent with large energy barriers, of order 20–30 T. In contrast, unzipping at a slow constant velocity is an *imperative* in the work of Bockelmann *et al.* [6–8]. Because these experiments are *effectively* at constant extension, the force in this case fluctuates as needed about $F_c(T)$ to insure that all energy barriers are overcome.

Many of the pauses and local "two-level system" oscillations found by the Prentiss group coincide with those predicted for phage lambda DNA [42], using energy landscapes obtained from the base pairing and stacking energies of Santa Lucia *et al.* [43]. Overall, these experiments seem consistent with important aspects of the theory reviewed here, such as large energy barriers and sequence-specific pause points. We should stress that longer run times or higher temperatures are necessary to obtain true thermodynamic equilibrium. Figure 12b shows the unzipping history of four different beads at $F \simeq 20\ pN$. There are again jumps between common plateaus as well as oscillations suggesting nearly degenerate minima. Although 20 pN should be above the thermodynamic critical force $F_c(T)$, the DNA nevertheless fails to unzip completely on experimental time scales, presumably due to the ruggedness of "downhill" landscapes such as that in Fig. 10. Predictions such as Eq. (42) are only be applicable to systems in true thermodynamic equilibrium.

Experimental checks of phase diagrams such as that in Fig. 2 and the associated physical predictions would be of considerable interest. As mentioned in Sec. 3, the phase diagram itself can be used to directly measure the temperature-dependent free energy of the DNA duplex at zero force. Analysis of this free energy is simplified if a theory of the unzipped handles is available. Analytic theories require forces large enough to stretch out the handles and prevent hairpins created by self-hybridization. The self-hybridization of ssDNA (for $F \lesssim 10 \ pN$) indicated in the experiments of Maier *et al.* [44] could be reduced by introducing single stranded binding proteins to smooth out the heterogeneity of the handles.

The assumption (36) of short-range correlations in the sequence is an approximation valid only for simple "coding DNA" in bacteria or viruses. The introns or so-called "junk DNA" present in many eukaryotic organisms may be better described by [36]

$$\overline{\eta(m)\eta(m')} \sim \frac{1}{|m - m'|^{2-2\beta}}, \tag{56}$$

with $\beta > 1/2$. Although we are unaware of detailed calculations, a straightforward generalization of the argument leading to Eq. (42) now gives [9]

$$\overline{\langle m \rangle} \sim 1/f^{\frac{1}{1-\beta}}. \tag{57}$$

For $\beta \leq 1/2$, we expect that predictions for sequences with short-range correlations will apply. The exponent $\beta = 0.55$ seems to describe the energy correlations of some intron-rich sequences [36].

Ideas similar to those discussed here can also be applied to force-induced denaturation [45, 14] of the multiple stem-loop structures which define the secondary structure of complicated RNA enzymes. Here, however, strings of RNA hairpins are unzipped simultaneously and the discrete jumps and plateaus in $\langle m \rangle$ near the unzipping force are averaged out.

Acknowledgments

Virtually all the theoretical work described here is contained in the Ph.D. thesis of my former student, David Lubensky. I am fortunate to have such students. I am grateful as well to Claudia Danilowicz, Mara Prentiss and other experimenters for their determination to test some of the predictions sketched here. Finally, I would like to thank D. Branton, R. Bundschuh, T. Hwa, and Y. Kafri for helpful conversations on this material. This work was supported by the National Science Foundation, through grant DMR 0231631 and in part through the Harvard Materials Research Laboratory, via Grant DMR 0213805.

References

[1] For a brief review, see C. Bustamante, Z. Bryant and S.B. Smith, *Nature* **421**, 423 (2003).

[2] F. Crick, *Nature* **227**, 561 (1970).

[3] M. Rief, M. Gautel, F. Oesterhelt, J.M. Fernandez, and J. Gaub, *Science* **276**, 1109 (1997).

[4] T.R. Strick, V. Croquette, and D. Bensimon, *PNAS* **95**, 10,579 (1998).

[5] J. Liphardt, B. Onoa, S.B. Smith, I. Tinoco, Jr., and C. Bustamante, *Science* **292**, 733 (2001).

[6] U. Bockelmann, Bessevaz, and F. Heslot, *Phys. Rev. Lett.* **79**, 4489 (1997); *Phys. Rev. E***58**, 2386 (1998).

[7] B. Essevaz-Roulet, U. Bockelmann, and F. Heslot, *Proc. Natl. Acad. Sci. USA* **94**, 11935 (1997).

[8] U. Bockelmann, P. Thomen, B. Essevaz-Roulet, V. Viasnoff, and F. Heslot, *Biophys. J.* **82**, 1537 (2002).

[9] D.K. Lubensky and D.R. Nelson, *Phys. Rev. Lett.* **85**, 1572 (2000).

[10] S.M. Bhattacharjee, *J. Phys.* **A48**, L423 (2000).

[11] K.L. Sebastian, *Phys. Rev. E***62**, 1128 (2000).

[12] S. Cocco, R. Monasson, and J.F. Marko, *Proc. Natl. Acad. Sci. USA* **98**, 8608 (2001); *Phys. Rev. E***65**, 0141907 (2002).

[13] D. Morenduzzo, S. Bhattacharjee, S. Maritan, E. Orlandini, and F. Seno, *Phys. Rev. Lett.* **88**, 028102 (2002).

[14] D.K. Lubensky and D.R. Nelson, *Phys. Rev. E***65**, 031917 (2002).

[15] B. Alberts, D. Bray, J. Lewis, M. Ruff, K. Roberts, and J.D. Watson, *Molecular Biology of the Cell* (Garland, New York, 1994).

[16] P.G. deGennes, *Rev. Mod. Phys.* **57**, 827 (1985).

[17] D. Poland and H.A. Scheraga, *Theory of Helix-Coil Transitions in Biopolymers* (Academic, New York, 1970).

[18] F.W. Wiegel, in *Phase Transitions and Critical Phenomena*, Vol. 7, edited by C. Domb and J. L. Lebowitz (Academic, New York, 1983).

[19] A.Y. Grossberg and A.R. Khokhlov, *Statistical Physics of Maromolecules* (AIP Press, New York, 1994), Chapter 7.

[20] In contrast to the finite kink energy, the entropy of an isolated kink diverges logarithmically with system size. See L.D. Landau and E.M. Lifshitz, *Statistical Physics* (Pergamon Press, Oxford, 1988), Sec. 163.

[21] M.E. Fisher, *J. Chem. Phys.* **45**, 1469 (1966).

[22] Y. Kafri, D. Mukamel, and L. Peliti, *Phys. Rev. Lett.* **85**, 4988 (2000); *Eur. Phys. J.* **B27**, 132 (2002).

[23] Y.L. Lyubchenko, A.V. Volodskii, and M. Frank-Kamenetskii, *Nature* **271**, 28 (1978); A.V. Volodskii, A.V. Amrykyan, Y.L. Lyubchenko, and M. Mamenetskii, *Biomol. Struct. Dyn.* **2**, 131 (1984).

[24] L.-H. Tang and H. Chate, *Phys. Rev. Lett.* **86**, 830 (2001); see also D. Cule and T. Hwa, *Phys. Rev. Lett.* **79**, 2375 (1997).

[25] M. Doi and S.F. Edwards, *The Theory of Polymer Dynamics* (Oxford, New York, 1988).

[26] P. Nelson, *Biological Physics: Energy, Information Life* (W.H. Freeman, New York 2003).

[27] P.G. deGennes, *Rep. Prog. Phys.* **32**, 187 (1969).

[28] See, e.g., J. Marko and E.D. Siggia, *Science* **265**, 506 (1994) for a discussion of these effects, which are related to the physics of supercoiling.

[29] For an analysis of the closely related problem of a vortex line in a Type II superconductor interacting with a columnar pin, see D.R. Nelson *Defects and Geometry in Condensed Matter Physics* (Cambridge University Press, Cambridge, 2002), Chapters 7 and 8.

[30] See, e.g., F. Schwabl, *Quantum Mechanics* (Springer, Berlin, 1992), Chapter 6.

[31] For potentials like those in Fig. 4, the ground state always remains localized in one and two dimensions provided the well depth $U_0 > 0$. See L.D. Landau and E.M. Lifshitz, *Quantum Mechanics* (Pergamon Press, New York, 1969).

[32] N. Hatano and D.R. Nelson, *Phys. Rev.* **B56**, 8651 (1997), and references therein.

[33] D. Ertas, *Phys. Rev.* **B59**, 188 (1999).

[34] J.F. Marko and E.D. Siggia, *Macromolecules* **28**, 8759 (1995).

[35] S.B. Smith, Y. Cui, and C. Bustamante, *Science* **271**, 795 (1996).

[36] S.V. Buldyrev, A.L. Goldberger, S. Havlin, R.N. Mantegna, M.E. Matsa, C.-K. Peng, M. Simons, and H.E. Stanley, *Phys. Rev.* **E51**, 5084 (1995).

[37] P. LeDoussal, C. Monthus, and D.S. Fisher, *Phys. Rev.* **E59**, 4795 (1999).

[38] Ya. G. Sinai, *Theor. Probab. Appl.* **27**, 247 (1982).

[39] B. Derrida, *J. Stat. Phys.* **31**, 433 (1983).

[40] J.P. Bouchaud, A. Comtet, A. Georges, and P. LeDoussal, *Ann. Phys. (N.Y.)* **201**, 285 (1990).

[41] A.B. Harris, *J. Phys. C* **7**, 1671 (1974).

[42] C. Danilowicz, V.W. Coljee, C. Bouzignes, D.K. Lubensky, D.R. Nelson, and M. Prentiss. *Proc. Nat'l. Acad. Science* **100**, 1694 (2003).

[43] J. Santa Lucia, H.T. Alawi and P.A. Seneviratne, *Biochemistry* **35**, 3555 (1996).

[44] B. Maier, D. Bensimon and V. Croquette, *PNAS* **97**, 12002 (2000).

[45] U. Gerland, R. Bundschuh and T. Hwa, *Biophys. J.* **81**, 1324 (2001).

CAN THEORY PREDICT TWO-STATE PROTEIN FOLDING RATES? AN EXPERIMENTAL PERSPECTIVE.

BLAKE GILLESPIE AND KEVIN W. PLAXCO
Department of Chemistry and Biochemistry
University of California, Santa Barbara
Santa Barbara, CA 93106

Abstract

The fastest simple, kinetically two-state protein folds a million times more rapidly than the slowest. Here we review theoretical models of protein folding kinetics in terms of their ability to qualitatively rationalize, if not quantitatively predict, this most basic of experimental observations. We find that the properties known to account for variations in the folding rates of simple on- and off-lattice computational models, such as the roughness of the energy landscape, the magnitude of the energy gap, or the equilibrium collapse parameter σ, do not account for the vast range of two-state folding rates observed in the laboratory. Similarly, two-state folding rates appear to be only modestly related to the structural details of a protein's "folding nucleus." Instead, because the equilibrium folding of two-state proteins is highly cooperative (perhaps accounting for their smooth landscapes, large energy gaps and low σ), it appears that their million-fold range of folding rates is predominantly defined by difficulty of the large-scale diffusive search required to find the native topology.

1. Introduction

Simple, single domain proteins fold via a process that is well-approximated as two-state, in which discrete, well-populated misfolded or partially folded states play no significant role in defining their folding rates [reviewed in 1]. Despite the potentially simplifying lack of intermediates, however, the folding rates of the dozens of kinetically two-state proteins characterized to date span a million-fold range (Fig. 1) [*e.g.* 2,3]. When coupled with the appealing simplicity of two-state behavior, this broad diversity of folding rates provides a uniquely straightforward and quantitative opportunity to test theories of the folding process.

Generally speaking, theoretical models of protein folding kinetics can be grouped into two classes. Perhaps the dominant class is comprised of theories derived from

A.T. Skjeltorp and A.V. Belushkin (eds.), Forces, Growth and Form in Soft Condensed Matter: At the Interface between Physics and Biology, 93-111.
© 2004 *Kluwer Academic Publishers. Printed in the Netherlands.*

observations of the simulated folding of simple on- and off-lattice computational models. The smaller class consists of theoretical models of folding which have emerged from observations of protein folding in the laboratory. In this article we broadly review recent simulation- and experiment-based theories of folding kinetics and critically evaluate these theories in terms of their ability to qualitatively rationalize, if not quantitatively predict, the vast range of folding rates observed for two-state proteins.

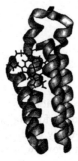

muscle acylphosphatase

Figure 1. The simple, single domain proteins cytochrome b_{562} and muscle acylphosphatase are of similar size and stability, yet while the former folds in microseconds, the latter folds in seconds [2,3]. Here we critically review the ability of various theories of protein folding to rationalize qualitatively, if not predict quantitatively, this million-fold range of rates.

2. Simulation-based theories of protein folding kinetics

The majority of contemporary theories of protein folding kinetics are based on observations of the simulated folding of simple, computational models. In an ideal world, the relevant simulations would entail a level of detail commensurate with the complex, atomistic structure of a fully solvated polypeptide. In reality, however, the simulation of protein folding in atomistic detail has proven computationally overwhelming. The difficulty is two fold. First, computational times scale strongly with the number of atoms and even the smallest proteins are comprised of thousands of atoms solvated by thousands of water molecules. Second, even the most rapidly folding proteins fold extremely slowly relative to the femtosecond time step of fully detailed MD simulations (for a recent solution to this dilemma, see Pande and co-workers folding@home project [*e.g.* 4]). Faced with this computational obstacle the large majority of the theoretical literature in protein folding has been based on observations not of fully detailed protein models, but on highly simplified representations of the polypeptide chain. Lattice polymers, perhaps the most popular computational model, simplify the description of the polypeptide chain by distilling each amino acid into a single "bead," and simplify folding dynamics by limiting the moves of each bead to hops between discrete points on a coarse lattice. While these lattice polymers are highly simplified, they do capture many of the relevant aspects of real proteins. For example, like proteins, lattice polymers are sequence-specific heteropolymers that can encode a unique native fold and, while the coarse lattice significantly reduces the entropy of folding, this entropy is still sufficiently large that folding would be slow were it a fully random search process. Similarly, like proteins, different lattice polymer sequences fold with vastly different rates. To date a major goal of computational folding studies has been the identification of the equilibrium properties that uniquely identify those rare lattice polymer sequences that fold rapidly, under the assumption that similar behavior will underlie the rapid folding of real proteins. The criteria predicted to distinguish between the rapidly and slowly folding lattice polymer sequences thus provide a clear opportunity for evaluating the correspondence between theory and experiment in protein folding.

The 1990's alone saw the publication of several hundred papers on the folding kinetics of simple on- and off-lattice protein models. While it is impossible to accurately distill such a large, diverse and sometimes conflicting literature into a few brief conclusions, much of this literature can be classified by the criterion that separates rapidly folding sequences from slowly folding sequences. Here we briefly describe the three major criteria that have been suggested as potential determinants of the relative folding rates of simplified computational models and review the experimental literature for clues as to whether similar criteria are responsible for the broad range of rates observed for the folding of two-state proteins.

2.1 SMOOTH ENERGY LANDSCAPES

A large body of theoretical work suggests that the rapid folding of lattice polymers [*e.g.* 5-7] and simplified off-lattice polymers [*e.g.* 8,9] is associated with smooth, "funnel-shaped" energy landscapes. That is, rapid folding will occur when the energy of each unfolded or partially folded conformation decreases more-or-less monotonically as conformations become more native-like along some reaction coordinate(s) (Fig. 2). If this energetic guidance does not occur, or if the landscape is rough (contains local minima deeper than a few k_BT that act as traps) folding becomes "glassy" and dominated by multiple kinetically trapped, misfolded states and slows dramatically. Differences in the roughness of the energy landscape can lead to orders of magnitude changes in the folding rates of simplified computational models [*e.g.* 7].

The contribution of energetic roughness to protein folding kinetics can be determined experimentally because roughness is a relative term; roughness only contributes to kinetics when it is large relative to k_BT. As T is lowered and energetic roughness begins to dominate the energy landscape, heterogeneous, non-single-exponential kinetics will arise [*e.g.* 5,7,10]. That is, as T decreases towards the "glass transition temperature" (T_g) the myriad of small, local barriers loom larger relative to k_BT and begin to retard the folding process. When this occurs, the kinetics will switch from single exponential $(h = 1)$ to a slower, stretched exponential $(h > 1)$:

$$S(t) = S_n + A_0\exp(-(k_f \cdot t)^{1/h}) \qquad (1)$$

where k_f is the folding rate, A_0 denotes the amplitude change upon folding, and S_n and $S(t)$ respectively denote the signal of the native state and that observed at time t [e.g. 5,7-10]. Using this method to measure the energetic roughness of both rapidly and slowly folding proteins, we can determine the contribution of roughness to relative folding rates.

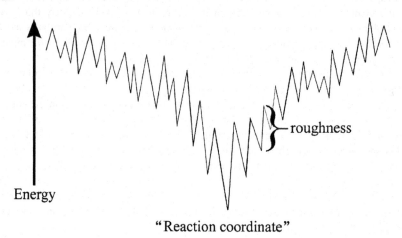

Figure 2. Wolynes, Onuchic and others have noted that rapid folding will occur when the energy of each unfolded or partially folded conformation decreases more-or-less monotonically (along some simple, but often difficult to define and multi-dimensional, order parameter) as conformations become more "native-like" [e.g. 5,6]. Roughness on this energy landscape produces slow, stretched-exponential kinetics for some lattice polymer sequences [e.g. 7] and may also account for part of the broad range of observed two-state folding rates

Do differences in the roughness of the energy landscape account for the different folding rates observed for two-state proteins? Non-single-exponential kinetics have been reported for the folding of the single domain protein ubiquitin under some conditions [11], suggesting that energetic roughness may contribute to relative folding rates. This work is complicated, however, by the presence of three proline residues in ubiquitin; due to significant *cis-trans* isomerization of the prolyl peptide bond, prolines can produce heterogeneities unrelated to landscape roughness. More recently we have monitored the folding kinetics of the simple, proline-free protein, protein L, at the lowest temperatures we can achieve in the laboratory [12]. An important caveat is that, unlike the situation *in silico*, in the laboratory proteins only fold in liquid water and thus T cannot be lowered arbitrarily. Using antifreezes we can probe folding rates at temperatures as low as −15°C; while this temperature is not very low on the absolute scale, it is more than 50°C below the temperature at which this protein evolved and thus well below the temperature at which nature presumably optimizes its energy landscape. Even at this temperature, however, we see no statistically significant evidence for heterogeneous, stretched exponential kinetics. Thus it appears that protein L's folding energy landscape is exceedingly smooth (T_g is unmeasurably low), and energetic roughness plays no significant role in defining its folding rate at physiological temperatures.

Of course, observing that the energy landscape of a single, relatively rapidly folding protein does not necessarily tell us much about the contributions that energetic roughness may play in defining relative folding rates. Therefore we have extended our observations to one of the most slowly folding two-state proteins reported to date [13], the pI3K SH3 domain. We find, however, that the slow folding of this molecule is also not a consequence of a rough energy landscape; its folding kinetics are well-fitted by a single exponential at the lowest temperatures we have probed [BG and KWP, manuscript in population]. It thus appears that, while theory is correct in associating smooth energy landscapes with the rapid folding of naturally occurring [e.g. 5], differences in the roughness of the energy landscape do not play a significant role in defining the million-fold range of observed two-state folding rates.

2.2 THE ENERGY GAP HYPOTHESIS

Karplus, Shakhnovich and co-workers have reported that a "necessary and sufficient criterion to ensure [rapid] folding is that the ground state be a pronounced energy minimum" relative to all other maximally compact states (Fig. 3) [14]. More recently these and other researchers have identified a number of related measures of this energy gap. These include the Z-score, which is the gap between the energy of the native state and the mean energy of all other maximally compact states relative to the standard deviation of the energy of the maximally compact states [15]. This latter measure of the energy gap is significantly correlated with the folding rates of a number of simple lattice polymers and is postulated to account for the vast dispersion observed in the folding rates of real proteins [16].

The energy gaps of almost all proteins are large. We know this because structural studies (NMR, crystallography, etc.) indicate that almost all proteins fold to a unique compact structure (the native state), and thus all other compact states must be significantly higher in energy than this ground state. Unfortunately, however, while we know the gap is large we cannot measure its precise magnitude and thus it is impossible to directly determine whether differences in the energy gap account wide range of observed protein folding rates. The difficulty arises because the gap in question is a gap in the "energy" between maximally compact states [14], and since the native state is the only maximally compact state observed experimentally, we cannot determine how stable it is relative to the other maximally compact states. Thus the energy gap cannot be measured in the laboratory in any of its incarnations using any experimental approach.

98

Energy

Figure 3. Karplus, Shakhnovich and others argue that, among conformations with a spectrum of energies, a large gap between the energy of the native state and the second lowest energy maximally compact state is a "necessary and sufficient condition for [rapid] folding" [*e.g.* 14] and that the size of this gap is correlated with folding rates [*e.g.* 16]. While it is not possible to test this hypothesis directly (because the energy gap does not correspond to any experimental observable –see text), an indirect test of the hypothesis is provided by the empirical observation that the melting temperature (T_m) of lattice polymers is highly correlated with both the energy gap and with their folding rates [16].

An indirect experimental test of the energy gap hypothesis may be provided, however, by the *empirical* observation that, for some lattice polymer models, the equilibrium melting temperature (T_m, often denoted T_f in the theoretical literature) is correlated with the size of the energy gap, leading to the prediction that folding rates should be correlated with T_m [16]. To the extent that this has been monitored experimentally, however, the folding rates of distinct proteins are generally not found to correlate with T_m. The folding rates and T_m of the naturally occurring and two highly modified variants of the engrailed homeodomain provide an extreme example; while the T_m of these three proteins range from 54°C to 116°C (one of the largest ranges reported for a set of related, single-domain proteins), their folding rates are identical to within experimental error [17,18, BG and KWP, unpublished data]. Similarly, there is no correlation between T_m and folding rates across mesophile-, thermophile-, and hyperthermophile-derived CspBs [19]. Thus is appears that, even for closely related proteins, the energy gap is not a significant determinant of relative folding rates.

A second indirect test of the energy gap hypothesis is provided by the observation that rapidly folding lattice polymer sequences (*i.e.* those with a pronounced energy gap) are uncommon. For example, simulations of randomly selected 27-mer lattice polymers indicate that only 3% to 15% of such sequences encode a sufficiently large energy gap to ensure rapid folding [14,20], and the fraction of rapidly folding sequences may be yet smaller for 125-mer lattice systems [21]. This predicted paucity of rapidly folding sequences is further supported by studies suggesting that rapidly folding sequences will be produced only rarely in the absence of selective pressures aimed at ensuring rapid folding [e.g. 22-24; but see also 25]. In order to test the postulate that the energy gap – and thus rapid folding- are only rarely produced in the absence of kinetic selections we have recently determined the folding kinetics of several of *de novo* designed proteins. Because these proteins were produced by design algorithms that consider only the equilibrium physics of the native state [26], and make no consideration at all about folding kinetics, they provide a unique opportunity to test the postulated rarity of rapidly folding sequences. In contrast to this theoretical prediction, however, the *de novo* designed proteins characterized to date fold with rates that are indistinguishable from that of the analogous naturally occurring protein [18]. Indeed, like their naturally occurring counterpart, the designed proteins characterized to date are among the most rapidly folding proteins reported to date.

2.3 COLLAPSE COOPERATIVITY

Thirumalai and co-workers have demonstrated that the relative folding rates of many simple on- and off-lattice polymer models are defined by a dimensionless, equilibrium parameter termed σ (Fig. 4). σ is a measure of the ease with which the denatured state undergoes non-specific, "Flory-type" coil-to-globule collapse relative to the ease with which the protein folds as solvent quality (*e.g.* temperature or denaturant concentration) is reduced [27-32]. Two equivalent measures of σ can be derived. The first, σ_T, is defined in terms of thermal unfolding and is related to the melting (T_m) and collapse (T_θ) temperatures by:

$$\sigma_T = 1 - T_m/T_\theta \qquad (2)$$

where T_m is the temperature (in Kelvins) at which half of a population of polymers is folded and T_θ is the temperature at which the geometric average R_g of the ensemble is midway between those of the folded and unfolded states. The second, σ_D, is reported to be an equivalent parameter obtained via chemical denaturation experiments [30; D. Thirumalai and D. Klimov, pers. com.]. It is defined by:

$$\sigma_D = 1 - C_m/C_\theta \qquad (3)$$

where C_m is the denaturant concentration at which half of the population of polymers is folded and C_θ is the denaturant concentration at which the average R_g of the ensemble is midway between those of the folded and unfolded states.

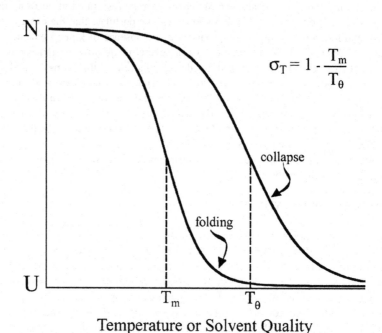

$$\sigma_T = 1 - \frac{T_m}{T_\theta}$$

Temperature or Solvent Quality

Figure 4. Thirumalai and co-workers have reported that the dimensionless, equilibrium collapse parameter σ is highly correlated with the relative folding rates of simple on- and off-lattice polymers [*e.g* 32]. σ is a measure of the point at which the unfolded polymer undergoes nonspecific coil-to-globule collapse relative to the point at with which it folds as the solvent quality or temperature is reduced. Shown is σ$_T$, which is defined in terms of thermal unfolding. An equivalent parameter, σ$_D$, is obtained via chemical denaturation.

Two research groups have explored the relationship between σ and the folding kinetics of simplified computational models. Thirumalai and co-workers have demonstrated that on- and off-lattice polymer sequences with high σ fold orders of magnitude more slowly than sequences that contract only at conditions nearer the midpoint of the folding transition [see, *e.g.*, 32]. Shakhnovich, Karplus and co-workers have also investigated the relationship between σ and folding rates [16]. While these authors report that folding kinetics correlate more strongly with the energy gap criterion than with σ (and, indeed, raise questions about the definition of the parameter) they nevertheless observe a statistically significant correlation between σ and folding rates across several lattice models. Simulations suggest that million-fold differences in folding rate should correspond to changes in σ of more than 0.5 [see, *e.g.*, 28,32,33], a difference well within reach of experimental verification.

Because thermal unfolding is often irreversible, σ_D is much more readily accessible to experiment than σ_T. Using the C_m and C_θ values obtained via CD and SAXS respectively, we have determined σ_D for four simple, single-domain [34]. All four exhibit σ_D effectively indistinguishable from zero. Similarly, we can employ previously reported data on the thermal unfolding of cytochrome c to determine that, for that protein, σ_T is also ~0 [35]. In keeping with theory, near zero values of σ are associated with the rapid folding of naturally occurring, two-state proteins. In contrast to the behavior of lattice and simple off-lattice polymers, however, the relative folding rates of simple proteins are not defined by this equilibrium measure of collapse cooperativity.

2.4 LATTICE-DERIVED CRITERIA FOR RAPID FOLDING ARE ALL OPTIMAL FOR PROTEINS

Experimental studies indicate that the criteria that determine folding rates in simple computational models do not account for the range of folding rates observed for simple proteins. Indeed, the energy gaps of nearly all simple proteins are immeasurably large and both energetic roughness and σ adopt values that are associated only with the most rapidly folding lattice polymers. Thus it appears that, while biologically relevant folding rates are associated with a smooth landscape/large energy gap/low σ, differences in these parameters do not account for the million-fold range of two-state folding rates observed in the laboratory.

3. Experiment-based theories of protein folding kinetics

In contrast to model-based theories of protein folding, several theories have arisen more directly from the experimental study of folding kinetics. Two of these, nucleation-condensation and the topomer search model, specifically address the issue of why some proteins fold more rapidly than others. Here we describe these two models in detail, placing emphasis on their ability to explain or predict the vast range of observed folding rates.

3.1. NUCLEATION-CONDENSATION

It has been surmised that native-like interactions are formed in the rate-limiting step of folding. For example, the perfectly exponential denaturant-dependencies of folding rates demonstrate that the folding transition-state contains interactions similar to those that stabilize the native state [reviewed in 36]. The role native interactions play in defining folding rates is further supported by reports that native-state stability is an important determinant of the relative folding rates of topologically similar proteins [37,38].

Protein engineering studies, termed φ-value analysis, have further clarified the nature of this native-like transition state structure, and led to the development of the so-called "nucleation-condensation" model of folding (Fig. 5). Many groups have noted that a sub-set of all mutations destabilize the folding transition state as much as they destabilize the native state, suggesting that the mutated residues are in a near-native environment during the rate-limiting step in folding [reviewed in 39]. While these structured transition-state residues are often distant in the protein sequence, they usually cluster when mapped onto the native fold [*e.g.* 40,41]. Taken together, these observations support the hypothesis that folding is akin to nucleation in a phase transition. That is, that the rate-limiting step in folding is the formation of a small, highly specific "nucleus" of native structure upon which the remaining structure condenses.

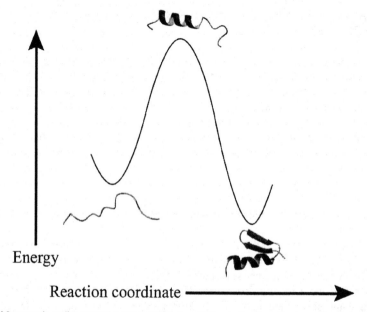

Figure 5. Mutagenesis studies (termed φ-value analysis [e.g. 39]) and other, less direct experimental evidence (see text), have lead to the suggestion that the formation of a small "nucleus" of native structure is sufficient to ensure productive folding. Differences in the stability of this element of native structure might thus account for the broad range of two-state folding rates.

This description of the rate-limiting step of folding is, however, qualitative; while such structures are clearly present in the transition state, their *quantitative* contribution to relative folding rates may be *limited*. Indeed, much recent evidence suggests that differences in the stability of the folding nucleus contribute relatively little to the million-fold range of two-state folding rates. This evidence includes the observation that the vast majority of point mutations [reviewed in 36] and even much larger sequence changes [18,42,43] produce less than order of magnitude changes in folding rates. Similarly, circular permutations [44,45] and covalent circularizations [46] that massively disrupt the native-like structure populated in the transition state produce only three to seven-fold changes in rate. Thus, while it is clear that a nucleus of native-like structure forms in the rate-limiting step in protein folding, differences in the structure –and presumably the thermodynamics- of this nucleus do not lead to significant variations in protein folding rate.

3.2 THE TOPOMER SEARCH MODEL

In contrast to experimental results suggesting that a small sub-set of native-state interactions dominate the thermodynamics of the folding transition state, the last five years have seen reports that several crude metrics of the global topology of the native state are highly correlated with two-state folding rates [36,47-50]. Recently, these empirical observations have been rationalized in terms of a quantitative, first-principles theory of folding termed the topomer search model.

The topomer search model postulates that relative barrier heights are dominated by the diffusive search for the set of unfolded conformations that share the native state's global topology [51]. Once this "native-topomer" is found, the rate-limiting step has been surmounted and specific native contacts rapidly form, generating the fully folded protein (Fig. 6) [51,52]. Two properties of simple, single domain proteins suggest that the topomer search dominates two-state folding rates. First, the formation of local structures such as helices, hairpins and loops is orders of magnitude more rapid than the rate-limiting step in folding [53-57]. Second, the folding free energy of such isolated structural elements, indeed, of almost all of the partially folded and misfolded states of single domain proteins, are near or above zero [see, *e.g.*, 58-60]. Because local structure formation is rapid, all of the conformers in the denatured ensemble will rapidly sample sequence-local elements of native structure (Fig. 6; transition B to A). For the vast majority of unfolded conformations, however, this zippering will stall when the formation of additional native structure would require large-scale, potentially slow rearrangement of the polypeptide chain (Fig. 6, state A). Because this partially folded state is unstable, it ruptures at least as rapidly as the rate with which it formed (Fig. 6, transition A to B), allowing the chain to diffuse into a new topology (Fig. 6, transition B to C). Only if the new topology is in the native topomer (Fig. 6, state C) can the formation of local structure proceed deep into the native well and the chain become trapped (Fig. 6, transition C to E). This model predicts that relative folding rates will be proportional to the probability of achieving the native topomer, and that differences in this probability account for the vast range of observed two-state folding rates.

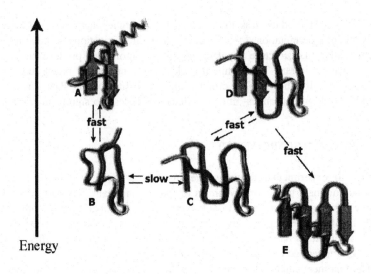

Energy

Figure 6. While sequence-local structures form far more rapidly than the rate-limiting step in folding, the extreme cooperativity of folding insures that these structures are almost always highly unstable and thus unproductive. It is only when the topology of the unfolded chain is near-native (*i.e.*, the chain is in the native topomer) that sufficient interactions can form to ensure productive folding.

Does the topomer search model *quantitatively* account for the wide range of two-state folding rates? Motivated by this question, we have recently developed a method of estimating the probability of an unfolded chain adopting any given topomer [52,61]. The method emerged from simulations of inert, Gaussian chains, which suggested that the probability of adopting a given topomer is quantifiable via a straightforward approximation. This approximation arises from two simplifying effects. First, because the locations of residues that are *close* in sequence are highly correlated, the probability of achieving a given topomer is dominated by pairs of residues that are *distant* in the sequence. The second is that the probability of ordering the chain is well described by a mean-field approximation: once a sufficient number of sequence-distant pairs of residues are brought into proximity, the probability of each of the remaining "ordering" events becomes independent of all other ordering events and approximately constant. The probability of forming the native topomer, $P(Q_D)$, is approximated by replacing the unique probability of ordering each specific pair with the *average* probability of ordering pairs [61]:

$$P(Q_D) \propto <K>^{Q_D} \tag{4}$$

where Q_D is the number of sequence-distant pairs whose proximity defines the topomer, $<K>$ is the average equilibrium constant for residue pairs being in proximity (and is less than unity). Folding rates should, in turn, be proportional $P(Q_D)$.

The prediction that folding rates relate to Q_D provides a means of testing the topomer search model. In order to perform this test, however, Q_D must be defined. The definition of Q_D assumes that any pair of residues that are separated by more than l_c residues along the sequence and are within distance r_c in the native state will be in proximity in the native topomer. The model is surprisingly insensitive to the precise values of these parameters; over a wide range of l_c and r_c, Q_D is correlated with $\log(k_f)$ with a correlation coefficient of $r^2 \sim 0.77$. Thus this simple model captures in excess of 3/4 of the variance in our kinetic data set using only the fitted parameters $<K>$ and a proportionality constant.

The predictive value of the topomer search model can be improved by introducing a length dependence. Polypeptide length is strongly correlated with the residuals of the simple model's predictions [52]. This correlation, however, suggests that, for two proteins with the same Q_D, the longer protein folds *more rapidly*. This observation, while statistically robust, is counter to the predictions of many simulations-based theories [*e.g.* 62-64]. With the addition of this inverse length dependence via the equation

$$k_f \propto Q_D \cdot J^{Q_D/N} \tag{5}$$

(where J is a constant < 1 that is analogous to $<K>$), the correlation coefficient for the topomer search model increases to $r^2 = 0.85$ [52]. The ability of this simple, near first-principles model of folding kinetics to capture 85% of the variance in two-state folding rates further supports the hypothesis that the topomer search process dominates relative barrier heights.

4. Why are the kinetics of lattice polymer folding different from that of proteins?

While a smooth landscape, a large energy gap or a sufficiently low σ may be necessary and sufficient conditions to ensure rapid heteropolymer folding, differences in these parameters do not account for differences in the folding rates of simple, single domain proteins. Instead the relative folding rates of simple proteins are defined by measures of their native state topology, apparently because the difficultly of finding the native topomer dominates relative barrier heights.

Why do the determinants of protein folding rates differ from those of lattice polymer models? One possibility is that energetic roughness (or a small energy gap or a low σ) overwhelms subtle, topology-dependent effects, and thus the topology-dependence of protein folding rates might only be apparent because their energy landscapes are very smooth. This argument suggests that, in the absence of energetic roughness, the folding rates of lattice polymers might similarly correlate with native-state topology. Because only native interactions are favorable in G♠ polymers [65], their energy landscapes are relatively smooth. G♠ polymers thus provide a means of testing this hypothesis. Extensive simulations of the folding of these models, however, demonstrate that smooth energy landscapes alone do not generate strongly topology-dependent folding rates; on-lattice G♠ polymer folding rates are effectively uncorrelated with topology [66,67] and the topology-dependence of more sophisticated, off-lattice G♠ polymers ranges from moderate [68] to non-existent [69]. No matter how smooth their energy landscape, none of the traditional polymer models exhibits the dramatic topology-dependence observed for two-state protein folding.

The topomer search model provides a potential rationale for the lack of topology-dependence. The topomer search dominates folding kinetics because the folding of small proteins is extremely cooperative. Since in excess of 90% of the native structure is required for the free energy of a typical single domain protein to drop below zero, only unfolded molecules in the native topomer can zipper enough contacts to fall into a stable well. In comparison to proteins, the folding of traditional G♠ polymers is relatively non-cooperative [70]. For example, the free energy of partially folded G♠ lattice polymers is a relatively linear function of the total number of interactions present [66], whereas the free energy of partially folded proteins falls precipitously only as the last few residues adopt their native conformation: the truncation of 1-3 residues from the structured termini of small, single domain proteins almost invariably leads to their complete unfolding [e.g., 58-60]. G♠ polymers can, however, be forced to adopt a degree of cooperativity by defining the energy of a given conformation as a non-linear function of the number of interactions present. Consistent with the predictions of the topomer search model, the introduction of such cooperativity leads to a highly significant relationship between topology and lattice-polymer folding rates [66,67]. If folding is cooperative and the energy landscape is smooth (i.e. lacking non-native traps), only those unfolded conformations in the native topomer can fold productively and thus the entropic cost of the topomer search will dominate relative folding rates.

Simulations thus support the suggestion that the topology-dependent folding rates observed for simple, single domain proteins is an unavoidable consequence of their highly cooperative folding. In addition to producing topology-dependent kinetics, however, the addition of cooperativity also *decelerates* the folding of G♠ polymers [66,67,70]. Does cooperativity also decelerate protein folding? We speculate that the net effect of cooperativity is to accelerate folding. Cooperativity provides a means of destabilizing partially structured, misfolded states relative to the native fold [66], and thus may accelerate folding rates more by destabilizing traps than it decelerates them by destabilizing potentially productive intermediates. This would suggest that the observed dominance of the topomer search in defining folding rates is a consequence of the cooperativity necessary to ensure the smooth energy landscape/large energy gap/low σ required to produce rapid folding.

5. Conclusions

The criteria that distinguish rapidly folding computational models from those that fold more slowly do not account for the broad range of rates observed for the folding of two-state proteins; the experimentally measurable criteria associated with rapidly folding lattice and off-lattice computational models appear to be perfectly optimized for many naturally occurring proteins. The rapid folding of *de novo* designed proteins also suggests that this optimization does not require explicitly kinetic selective pressures or design considerations. And while native-like interactions are clearly formed during the rate-limiting step in folding, they are apparently not a dominant contributor to relative barrier heights. Finally, a simple mathematical description of the diffusive search for the correct overall topology –the topomer search model- accurately predicts relative folding rates. The topomer search dominates folding of two-state proteins because their equilibrium folding is so cooperative that only unfolded conformations in the native topomer can zipper sufficient structure for the free energy to drop below zero. We speculate that this cooperativity provides a means of destabilizing partially structured, misfolded states relative to the native fold, thereby accelerating folding rates more by destabilizing traps (smoothing the landscape/improving the energy gap/decreasing σ) than it decelerates them by destabilizing potentially productive intermediates.

Acknowledgements

This work was supported in part by NIH grant R01GM62868-01A1 (KWP), BioSTAR grant s97-79 (KWP), and ACS junior postdoctoral research fellowship ACS CD INC 2-5-00 (BG).

References

1.   Jackson, S.E. (1998) How do small single domain proteins fold? *Fold. Des.* **3**, R81-91.

2.   vanNuland, N.A.J., Chiti, F., Taddei, N., Raugei, G., Ramponi, G., and Dobson, C.M. (1998) Slow folding of muscle acylphosphatase in the absence of intermediates. *J. Mol. Biol.* **283**, 883-891.

3.   Wittung-Stafshede, P., Lee, J.C., Winkler, J.R., and Gray, H.B. (1999) Cytochrome b_{562} folding triggered by electron transfer: Approaching the speed limit for formation of a four-helix bundle. *Proc. Natl. Acad. Sci. USA* **96**, 6587-6590.

4.   Zagrovic, B., Snow, C.D., Shirts, M.R., and Pande, V.S. (2002) Simulation of folding of a small alpha-helical protein in atomistic detail using worldwide-distributed computing. *J. Mol. Biol.* **323**, 927-937.

5.   Bryngelson, J.D. and Wolynes, P. G. (1987) Spin glasses and the statistical mechanics of protein folding. *Proc. Natl. Acad. Sci. USA* **84**, 7524-7528.

6.   Onuchic, J.N., Wolynes, P.G., Luthey-Schulten, Z., and Socci, N.D. (1995) Toward an outline of the topography of a realistic protein folding-funnel. *Proc. Natl. Acad. Sci. USA* **92**, 3626-3630.

7.   Onuchic, J.N., Luthey-Schulten, Z., and Wolynes, P.G. (1997) Theory of protein folding: The energy landscape perspective. *Ann. Rev. Phys. Chem.* **48**, 545-600.

108

8.   Thirumalai, D., Ashwin, V., and Bhattacharjee, J.K. (1996) Dynamics of random hydrophobic-hydrophilic copolymers with implications for protein folding. *Phys. Rev. Lett.* **77**, 5385-5388.

9.   Nymeyer, H., García, A.E., and Onuchic, J.N. (1998) Folding funnels and frustration in off-lattice minimalist protein landscapes. *Proc. Natl. Acad. Sci. USA* **95**, 5921-5928.

10. Socci, N.D., Onuchic, J.N., and Wolynes, P.G. (1998) Protein folding mechanisms and the multidimensional folding funnel. *Prot.: Struct. Func. Gen,* **32**, 136-158.

11. Sabelko, J., Ervin, J., and Gruebele, M. (1999) Observation of strange kinetics in protein folding. *Proc. Natl. Acad. Sci. USA* **96**, 6031-6036.

12. Gillespie, B. and Plaxco, K.W. (2000) Non-glassy kinetics in the folding of a simple, single domain protein. *Proc. Natl. Acad. Sci. USA* **97**, 12014-12019.

13. Guijarro, J.I., Morton, C. J., Plaxco, K.W., Campbell, I.D., and Dobson, C.M. (1998) Folding kinetics of the SH3 domain of PI3 by real-time NMR and optical techniques. *J. Mol. Biol.* **275**, 657-667.

14. Sali, A., Shakhnovich, E.I., and Karplus, M. (1994). How does a protein fold? *Nature* **369**, 248-251.

15. Gutin, A.M., Abkevich, V.I., and Shakhnovich, E.I. (1995) Evolution-like selection of fast-folding model proteins. *Proc. Natl. Acad. Sci. USA* **92**, 1282-1286.

16. Dinner, A.R., Abkevich, V., Shakhnovich, E.I., and Karplus, M. (1999) Factors that affect the folding ability of proteins. *Prot. Struct. Func. Gen.* **35**, 34-40.

17. Mayor, U., Johnson, C.M., Daggett, V., and Fersht, A.R. (2000) Protein folding and unfolding in microseconds to nanoseconds by experiment and simulation. *Proc. Natl. Acad. Sci. USA* **97**, 13518-13522.

18. Gillespie, B., Vu, D., Shah, P.S., Marshall, S., Dyer, R.B., Mayo, S.L., and Plaxco, K.W. (2003) NMR and Temperature Jump measurements of *de novo* designed proteins demonstrate rapid folding in the absence of explicit selection for kinetics. *J. Mol. Biol.* In press.

19. Perl, D., Welker, C., Schindler, T., Schroder, K,, Marahiel, M.A., Jaenicke, R., and Schmid, F.X. (1998) Conservation of rapid two-state folding in mesophilic, thermophilic and hyperthermophilic cold shock proteins. *Nat. Struct. Biol.* **5**, 229-235.

20. Shakhnovich, E., Fartzdinov, G., Gutin, A.M., and Karplus, M. (1991) Protein folding bottlenecks, *Phys. Rev. Lett.* **67**, 1665-1668.

21. Dinner, A.R., So, S., and Karplus, M. (1998) Use of quantitative structure-property relationship to predict the folding ability of model proteins. *Prot. Struct. Func. Gen.* **33**,177-203.

22. Mirny, L.A., Abkevich, V.I., and Shakhnovich, E.I. (1998) How evolution makes proteins fold quickly. *Proc. Natl. Acad. Sci. USA* **95**, 4976-4981.

23. Gutin, A., Sali, A., Abkevich, V.I., Karplus, M., and Shakhnovich, E.I. (1998) Tempertature dependence of the folding rate in a simple protein model: Search for a "glass" transition. *J. Chem. Phys.* **108**, 6466-6483.

24. Li, L., Mirny, L.A., and Shakhnovich, E.I. (2000) Kinetics, thermodynamics and evolution of non-native interactions in a protein folding nucleus. *Nat. Struct. Biol.* **7**, 336-342.

25. Larson, S., Ruczinski, I., Davidson, A.R., Baker, D., and Plaxco, K.W. (2002) Protein folding nuclei exhibit little evidence of preferential conservation. *J. Mol. Biol.* **316**, 225-233.

26. Dahiyat, B.I. and Mayo, S.L. (1996) Protein design automation. *Prot. Sci.* **5**, 895-903.

27. Camacho, C.J. and Thirumalai, D. (1996) A criterion that determines the foldability of proteins. *Europhys. Lett.* **35**, 627-632.

28. Klimov, D.K. and Thirumalai, D. (1996) Criterion that determines the foldability of proteins. *Phys. Rev. Lett.* **76**, 4070-4073.

29. Klimov, D.K and Thirumalai, D. (1996) Factors governing the foldability of proteins. *Prot. Struct. Func. Gen.* **26**, 411-441.

30. Klimov, D.K. and Thirumalai, D. (1998) Cooperativity in protein folding: from lattice models with sidechains to real proteins. *Fold. Des.* **3**, 127-139.

31. Klimov, D.K. and Thirumalai, D. (1998) Linking rates of folding in lattice models of proteins with underlying thermodynamic characteristics. *J. Chem. Phys.* **109**, 4119-4125.

32. Thirumalai, D. and Klimov, D.K. (1999) Deciphering the timescales and mechanisms of protein folding using minimal off-lattice models. *Curr. Opin. Struct. Biol.* **9**, 197-207.

33. Veitshans, T., Klimov, D.K., and Thirumalai, D. (1997) Protein folding kinetics: Timescales, pathways and energy landscapes in terms of sequence-dependent properties. *Fold. Des.* **2**, 1-22.

34. Millet, I.S., Townsley, L., Chiti, F., Doniach, S., and Plaxco, K.W. (2002) Equilibrium collapse and the kinetic 'foldability' of proteins. *Biochemistry* **41**, 321-325.

35. Hagihara, Y., Hoshino, M., Hamada, D., Kataoka, M., and Goto, Y. (1998) Chain-like conformation of heat-denatured ribonuclease A and cytochrome c as evidenced by solution X-ray scattering *Fold. Des.* **3**, 195-201.

36. Plaxco, K.W., Simons, K.T., Ruczinski, I., and Baker, D. (2000) Sequence, stability, topology and length; the determinants of two-state protein folding kinetics. *Biochemistry* **39**, 11177–11183.

37. Clarke, J., Cota, E., Fowler, S.B., and Hamill, S. J. (1999) Folding studies of immunoglobulin-like β-sandwich proteins suggest that they share a common folding pathway. *Fold. Des.* **7**, 1145-1153.

38. Plaxco, K.W., Guijarro J.I., Morton, C.J., Pitkeathly, M., Campbell, I.D., and Dobson, C.M. (1998) The folding kinetics and thermodynamics of the Fyn-SH3 domain. *Biochemistry* **39**, 2529-2537.

39. Fersht, A. R. (1994) Characterizing the transition states in protein folding: an essential step in the puzzle. *Curr. Opin. Struct. Biol.* **5**, 79-84.

40. Jackson, S.E., elMasry, N., and Fersht, A.R. (1993) Structure of the hydrophobic core in the transition-state for folding of chymotripsin inhibitor-2 – a critical test of the protein engineering method of analysis. *Biochemistry* **32**, 11270-11278.

41. Riddle, D.S., Grantcharova, V.P., Santiago, J.V., Alm, E., Ruczinski, I., and Baker, D. (1999) Experiment and theory highlight role of native state topology in SH3 folding. *Nat. Struct. Biol.* **6**, 1016-1024.

42. Riddle, D.S., Santiago, J.V., Bray-Hall, S.T., Doshi, N., Grantcharova, V.P., Yi, Q., and Baker, D. (1997) Functional rapidly folding proteins from simplified amino acid sequences. *Nat. Struct. Biol.* **4**, 805-809.

43. Kim, D.E., Gu, H., and Baker, D. (1998) The sequences of small proteins are not extensively optimized for rapid folding by natural selection. *Proc. Natl. Acad. Sci. USA* **95**, 4982-4986.

44. Viguera, A.R., Serrano, L., and Wilmanns, M. (1996) Different folding transition states may result in the same native structure. *Nat. Struct. Biol.* **3**, 874-880.

45. Lindberg, M., Tangrot, J., and Oliveberg, M. (2002) Complete change of the protein folding transition state upon circular permutation. *Nat. Struct. Biol.* **9**, 818-822.

46. Grantcharova, V.P., and Baker, D. (2001) Circularization changes the folding transition state of the src SH3 domain. *J. Mol. Biol.* **306**, 555-563.

47. Plaxco, K.W., Simons, K.T., and Baker, D. (1998) Contact order, transition state placement and the refolding rates of single domain proteins. *J. Mol. Biol.* **277**, 985-994.

48. Gromiha, M.M. and Selvaraj, S. (2001) Comparison between long-range interactions and contact order in determining the rate of two-state proteins: application of long-range order to folding rate prediction. *J. Mol. Biol.* **310**, 27-32.

49. Zhou, H. and Zhou, Y. (2002) Folding rate prediction using total contact distance. *Biophys. J.* **82**, 458-463.

50. Miller, E.J., Fischer, K.F., and Marqusee, S. (2002) Experimental evaluation of topological parameters determining protein-folding rates. *Proc. Natl. Acad. Sci. USA* **99**, 10359-10363.

51. Debe, D.A., Carlson, M.J., and Goddard, W.A. (1999) The topomer-sampling model of protein folding. *Proc. Natl. Acad. Sci. USA* **96**, 2596-2601.

52. Makarov, D.E. and Plaxco, K.W. (2003) The topomer search model: a quantitative description of two-state protein folding kinetics. *Prot. Sci.* **12**, 17-26.

53. Hagen, S.J., Hofrichter, J., Szabo, A., and Eaton, W.A. (1996) Diffusion-limited contact formation in unfolded cytochrome c: Estimating the maximum rate of protein folding. *Proc. Natl. Acad. Sci. USA* **93**,11615-11617.

54. Thompson, P.A., Eaton, W.A., and Hofrichter, J. (1997) Laser temperature jump study of the helix•coil kinetics of an alanine peptide interpreted with a 'kinetic zipper' model. *Biochemistry* **36**, 9200-9210.

55. Muñoz V., Thompson, P.A., Hofrichter, J., and Eaton, W.A. (1997) Folding dynamics and mechanism of beta-hairpin formation. *Nature* **390**,196-199.

56. Bieri, O., Wirz, J., Hellrung, B., Schutkowski, M., Drewello, M., and Kiefhaber, T. (1999) The speed limit for protein folding measured by triplet-triplet energy transfer. *Proc. Nat. Acad. Sci. USA* **96**, 9597-9601.

57. Lapidus, L.J., Eaton, W.A., and Hofrichter, J. (2000) Measuring the rate of intramolecular contact formation in polypeptides. *Proc. Natl. Acad. Sci. USA* **97**,7220-7225.

58. Flanagan, J.M., Kataoka, M., Shortle, D., and Engelman, D.M. (1992) Truncated staphylococcal nuclease is compact but disordered. *Proc. Natl. Acad. Sci. USA* **89**, 748-752.

59. Ladurner, A.G., Itzhaki, L.S., Gay, G.D., and Fersht, A.R. (1997) Complementation of peptide fragments of the single domain protein chymotrypsin inhibitor 2. *J. Mol. Biol.* **273**, 317-329.

60. Camarero, J.A., Fushman, D., Sato, S., Giriat, I., Cowburn, D., Raleigh, D.P., and Muir, T.W. (2001) Rescuing a destabilized protein fold through backbone cyclization. *J. Mol. Biol.* **308**, 1045-1062.

61. Makarov, D.E., Keller, C.A., Plaxco, K.W., and Metiu, H. (2002) How the folding rate constant of simple-single domain proteins depends on number of native contacts. *Proc. Natl. Acad. Sci. USA* **99**, 3535-3539.

62. Thirumalai, D. (1995) From minimal models to real proteins: time scales for proteins folding kinetics. *J. Phys. I France* **5**, 1457-1467.

63. Gutin, A.M., Abkevich, V.I., and Shakhnovich, E.I. (1996) Chain length scaling of protein folding time. *Phys. Rev. Lett.* **77**, 5433-5436.

64. Debe, D.A. and Goddard, W.A. (1999) First principles prediction of protein folding rates. *J. Mol. Biol.* **294**, 618-625.

65. Abe, H. and Gō, N. (1981) Noninteracting local-structure model of folding and unfolding transition in globular proteins. II. Application to two-dimensional lattice proteins. *Biopolymers* **20**, 10113-1031.

66. Jewett, A.I., Pande, V.S., and Plaxco, K.W. (2003) Cooperativity, smooth energy landscapes and the origins of topology-dependent protein folding rates. *J. Mol. Biol.* **326**, 247-253.

67. Kaya, H. and Chan, H.S. (2003) Contact order dependent protein folding rates: kinetic consequences of a cooperative interplay between favorable nonlocal interactions and local conformational preferences. *Proteins* in press.

68. Koga, N. and Takada, S. (2001). Roles of native topology and chain-length scaling in protein folding: A simulation study with a Gō-like model. *J. Mol. Biol.* **313**, 171-180.

69. Cieplak, M. and Hoang, T.X. (2003). Universality classes in folding times of proteins. *Biophys. J.* **84**, 475-488.

70. Eastwood, M.P. and Wolynes, P.G. (2001). Role of explicitly cooperative interactions in protein folding funnels: a simulation study. *J. Chem. Phys.* **114**, 4702-4716.

COPOLYMERS WITH LONG-RANGE CORRELATIONS: SEQUENCE DESIGN NEAR A SURFACE

N.Yu. STAROVOITOVA
Department of Physical Chemistry, Tver State University, Sadovy per. 35, 170002 Tver, Russia

P.G. KHALATUR
Department of Polymer Science, University of Ulm, Albert-Einstein-Allee 11, Ulm, D-89069, Germany

A.R. KHOKHLOV
Physics Department, Moscow State University, 119899 Moscow, Russia; Department of Polymer Science, University of Ulm, Albert-Einstein-Allee 11, Ulm, D-89069, Germany

Using Monte Carlo simulations and the lattice bond-fluctuation model, we perform the computer-aided sequence design of two-letter (AB) quasirandom copolymers with quenched primary structure near an infinite planar surface. To generate the conformation-dependent primary structures, two different approaches are employed. One of them is a variant of the "coloring" procedure (i.e., the conformation-dependent chemical modification of monomer units) introduced earlier by us for a solution of polymer globules. In this case, AB copolymers are prepared by adsorbing a bare homopolymer chain onto a flat substrate, after which its adsorbed segments are transformed to type A ones, and its unadsorbed segments to type B ones. The second approach represents an irreversible radical copolymerization process of selectively adsorbed A and B monomers with different affinity to a surface. We show that the statistical properties of the copolymer sequences obtained via the "coloring" procedure exhibit long-range correlations of the Levy-flight type similar to those known for proteinlike copolymers. On the other hand, copolymerization near the adsorbing surface leads to a copolymer with a specific quasi-gradient primary structure and well-pronounced long-range correlations in distribution of different monomer units along the chain. Some characteristics of the adsorbed single chains (statistics of trains, loops, and tails) are also studied.

A.T. Skjeltorp and A.V. Belushkin (eds.), Forces, Growth and Form in Soft Condensed Matter: At the Interface between Physics and Biology, 113-134.

1. Introduction

For a long time, chemical industry was interested in polymers mainly from the viewpoint of obtaining unique construction materials (plastics, rubbers, fibers, etc.). Couple of decades ago the main focus of interest shifted to functional polymers (superabsorbents, membranes, adhesives, etc.). In the nineties scientific and industrial polymer community started to discuss "smart" of "intellectual" polymer systems (e.g., soft manipulators, polymer systems for controlled drug release, field-responsive polymers); the meaning behind this term is that the functions performed by polymers become more sophisticated and diverse. This line of research concentrating on polymer systems with more and more complex functions will be certainly in the mainstream of polymer science in the 21st century.

One of the ways to obtain new polymers for sophisticated functions is connected with the synthesis of novel monomer units where the required function is linked to the chemical structure of these units. However, the potential of this approach is rather limited, because complicated and diverse functions of polymer material would then require a very complex structure of monomer units, which normally means that the organic synthesis is more expensive and less robust.

The alternative approach is to use known monomer units and to try to design a copolymer chain with given sequence of these units. There are practically infinite possibilities to vary sequences in copolymers: from the variation of some simple characteristics like composition of monomer units, average length of blocks (for the chains with blocky structure), availability of branching, etc. to more sophisticated features like long-range correlations or gradient structure. Therefore, in this approach a wide variety of new functional copolymers can be tailored.

It is important to emphasize that the nature has chosen this way in the evolution of main biological macromolecules: DNA, RNA, and proteins. These polymers in living systems are responsible for functions, which are incomparably more complex and diverse than the functions, which we are normally discussing for synthetic copolymers. The molecular basis for this ability to perform sophisticated functions is associated with unique primary sequences of units in biopolymers, which emerged in the course of biological evolution.

Thus, one of the promising approaches in the sequence design of functional copolymers is biomimetic in its nature: it is tempting to look at the main features of sequences of monomer units in biopolymers, understand how these sequences define functional properties, and then try to implement similar ideas for synthetic copolymers.

Some time ago, we started a detailed investigation of the methods of design of sequences of monomer units in copolymers with special emphasis on biomimetic approaches. In this way we hope to formulate new methods of synthesis of copolymers with sophisticated functional properties. Also, one of the motivations for this investigation is the attempt to understand, at least partially, the principles of evolution of sequences of biological macromolecules at the early stages of this evolution.

The first ideas connected with biomimetic sequence design of functional copolymers were formulated in 1998 [1-4]. They were based on the simple and well-known fact that the functioning of all globular proteins depends on two main factors: (i) they are globular; (ii) they are soluble in aqueous medium. It should be mentioned that the combination of these two factors is non-trivial, e.g., for homopolymers and random copolymers the transition to globular conformation is usually accompanied by the precipitation of globules from the solution [1,2]. Protein globules are soluble in water because of the special primary sequence: in the native conformation, most of hydrophobic monomer units are in the core of the globule while hydrophilic and charged monomer units form the envelope of this core. Of course, the division of 20 types of monomer units available in globular proteins in only two classes (hydrophobic and hydrophilic) is rather rough, but still this viewpoint gives a correct general picture of the structure of a protein. Now, having in mind the biomimetic approach described above, we can formulate the following problem: whether it is possible to design such sequence of synthetic AB copolymer (copolymer consisting of monomer units of two types, A and B) that in the most dense globular conformation all the hydrophobic B-units are in the core of this globule while hydrophilic A-units form the envelope of this core? This question was first addressed in Ref. [1] (see also Refs. [2-4]) and the corresponding polymers were called proteinlike AB copolymers.

The proteinlike AB sequences were first obtained in computer experiments [1-4], which can be described as follows. We start with arbitrary homopolymer globule conformation formed due to the strong attraction of monomer units [Fig. 1(a)] and perform for it a "coloring" procedure [Fig. 1(b)]: monomer units in the center of the globule are called A-type (hydrophobic) units, while monomer units belonging to globular surface are assigned to be B-type (hydrophilic) units. Then this primary structure is fixed, attraction of monomer units is removed and proteinlike copolymer is ready for the further investigation [Fig. 1(c)].

In particular, in Refs. [1-4] the coil-globule transition for thus generated AB copolymers was studied. This transition was induced by the attraction of only A-units (the interactions B-B and A-B were chosen to be repulsive). The properties of this transition were compared with those for random AB copolymers with the same composition and random-block AB copolymers with the same composition and the same "degree of blockiness" as for designed proteinlike AB copolymers. The calculations were performed by Monte-Carlo method using the bond-fluctuation model.

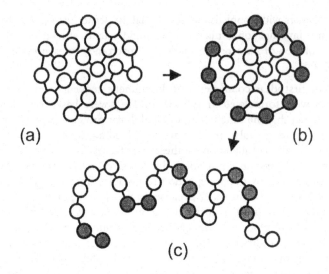

Figure 1. Main steps of the sequence design scheme for proteinlike copolymers: (a) homopolymer globule; (b) the same globule after "coloring" procedure; (c) proteinlike copolymer in the coil state.

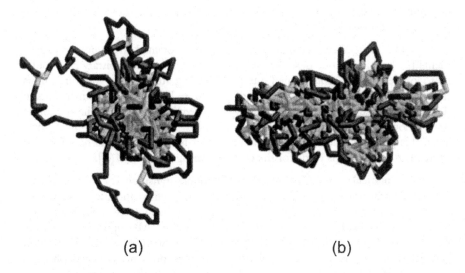

Figure 2. Typical snapshots of globular conformation for (a) proteinlike and (b) random copolymers.

It was shown [1-4] that the coil-globule transition in proteinlike copolymers occurs at higher temperatures, leads to the formation of denser globule, and has faster kinetics than for random and random-block counterparts. The reason for this is illustrated in Fig. 2 where the typical snapshots of globules formed by proteinlike and random AB copolymers with the same AB composition are shown. One can see that the core of proteinlike globule is much more compact and better formed, it is surrounded by the loops of hydrophilic units, which stabilize the core. Apparently, this is due to some memory effect: the core which existed in the "parent" conformation [this is the term introduced in Ref. [1] to describe the conformation of Fig. 1(b) where the coloring is performed] was simply reproduced upon refolding caused by the attraction of A units. One may say that the features of "parent" conformation are "inherited" by the proteinlike ABcopolymer. Looking at the conformations of Fig. 2, it is natural to argue that proteinlike copolymer globule should be soluble in water (see also Refs. [7,8]) and thus open to further modification in the course of biological evolution, while random copolymer globules will most probably precipitate and thus drop out of the evolution.

Returning to the computer-generated proteinlike sequences (Fig. 1), it is clear that they should exhibit long-range correlations along the chain, since the type of monomer unit (A or B) depends on the conformation of globule as a whole, not on the properties of some small part of the chain. In Refs. [9,10] it was shown, both by exact analytical theory and by computer simulation, that this is indeed the case and that the long-range correlations in the proteinlike sequences can be described by the so-called Levy-flight statistics [11].

After the idea of sequence design of proteinlike copolymers was presented and realized in computer simulations, several teams started experimental research aimed to obtain such copolymers in synthetic chemical laboratory.

In particular, Prof. H. Tenhu at the University of Helsinki studied grafting of short poly(ethylene oxide) (PEO) chains to the copolymer of thermosensitive N-isopropylacrylamide (NIPA) and glicydil methacrylate [12,13]. At room temperatures, such copolymer is in the coil state and grafting takes place in a random manner. At elevated temperatures, the transition to globule occurs, and grafting proceeds mainly in the globular surface, thus leading to its hydrophilization and to the creation of proteinlike copolymer in the sense described above. Indeed, it was shown that proteinlike copolymer prepared in this way exhibits solution turbidity at higher temperatures than the random one, and gives smaller aggregates in the turbid solution.

In the group of Prof. V. Lozinsky (Institute of Organoelement Compounds of Russian Academy of Sciences) the redox-initiated free-radical copolymerisation of thermosensitive N-vinylcaprolactam (NVCa) with hydrophilic N-vinylimidazole (NVIz) was studied at different temperatures [14]. At room temperatures, such polymerization gives a random copolymer. On the other hand, when polymerization takes place at elevated temperatures (ca. 65°C) growing chains form globules, and the concentration of monomers around the active radical is influenced by this fact. The conditions were found when proteinlike copolymers are emerging as a result of such synthesis. These copolymers were not precipitating at all when the solution is heated up to 80°C; on the other hand, dense globules were formed already around 30°C.

In the group of Prof. B. Mattiasson (the University of Lund, Sweden) similar type of proteinlike copolymers were obtained for the pair of monomers NIPA/NVIAz synthesized in aqueous solution [15]. It was demonstrated that the copolymer with virtually random distribution of NVIAz units along the chains did not interact with the metal chelate adsorbent, Cu^{2+}-iminodiacetate-Sepharose, whereas the copolymer, possessing the proteinlike sequence, was absorbed specifically by the resin, since the hydrophilic pendant imidazole groups were accumulated in the outer hydrophilic shell of macromolecular coil.

Recently, we performed computer experiments specially designed to describe the process of copolymerization with simultaneous globule formation in order to give more careful theoretical foundation for this synthetic method proposed in Ref. [14]. We have shown that such copolymerization process does indeed lead to the formation of a globule with proteinlike sequences [16].

The idea of conformation-dependent sequence design can be generalized. Indeed, the special primary sequence can be obtained not only from globular conformation; any specific polymer chain conformation can play the role of a parent one. The simplest example of this kind is connected with the conformation of a homopolymer chain adsorbed on a flat substrate. Let us "color" the chain segments being in direct contact with the surface in some typical instant conformation. This corresponds to the assumption that the surface catalyses some chemical transformation of the adsorbed segments. Then we will end up with an AB copolymer for which the sequence design was performed in the parent (adsorbed) state. One can expect that, after desorption, such an AB copolymer will have special functional properties: it will be "tuned to adsorption". Indeed, computer simulations [17] performed along this line have shown that, similar to proteinlike copolymers, "adsorption-turned copolymers" (ATC's) exhibit some special properties; for example, they adsorb better than their random and random-block counterparts under the same conditions [17]. In other words, the resulting copolymer sequence memorizes the original state of adsorbed homopolymer chain.

Pursuing this line further, we will focus in the present paper on the computer-aided synthesis of amphiphilic quasirandom copolymers with quenched primary structure near a macroscopic planar substrate. To this end, we will use the following two routes. (i) On the basis of an MC simulation technique, we will imitate the chemical modification ("coloring") of a partially adsorbed homopolymer chain. This surface-induced conformation-dependent sequence design is similar to that described above for bare polymer globules dissolved in a bulk solution and has been earlier realized for obtaining "adsorption-turned copolymers". Here, our main aim is to analyze in detail the statistics of arising copolymers for sufficiently long chains. (ii) The second approach is aimed at computer synthesis of copolymers from two types of selectively adsorbed monomers A and B distributed near a planar substrate with given gradients. We will simulate the irreversible growth process in the dilute regime for polymerizing monomers with different affinity to the surface. The polymerization will be modeled as step-by-step chemical reaction of addition of A and B monomer units to the growing copolymer chain, assuming that depolymerization reaction is not allowed in the simulation. Then, we will investigate the adsorption and statistical properties of prepared sequences to compare with those known for other types of copolymers (random, random-block, etc.).

2. Computational Technique and Models

For the sake of computational efficiency, we do not take into account here the chemical details of simulated polymer chains, addressing the generic features of surface-induced sequence formation only. Thus, all simulations are performed with the standard bond-fluctuation model, which is described in detail in Ref. [18]. This is a fast lattice algorithm that in three dimensions ($3d$) offers a set of 108 bond vectors with five different lengths. These are chosen such that the local self-avoidance prevents chains from crossing each other during their motion. The fluctuating bond length permits the implementation of random, local monomer displacements such that the dynamics of an isolated chain can be well-described by the Rouse model. In contrast to usual lattice algorithms, one monomer does not occupy one lattice site but eight sites in three dimensions. The motion of monomer units is generated by the Monte Carlo (MC) technique: A monomer unit and a lattice direction are chosen at random, and a move is attempted in the proposed direction. The move is successful, if the targeted sites are empty (excluded volume interaction), if it allows to be compatible with chain connectivity, and if it lowers the potential energy. On the other hand, if the energy difference, ΔU, between the final and the initial state is positive, the move is only accepted with probability $\exp[-\Delta U/T]$ according to the standard Metropolis algorithm [19], where T denotes the temperature (Boltzmann's constant $k_B = 1$). By means of this probability, temperature is introduced into the simulation. The chains are placed in a three-dimensional cubic lattice, which is infinite in the x and y directions and limited in the z direction by the two planes. In our computer experiment only excluded volume interactions between monomer units were taken onto account. Thus, thermodynamically good solvent conditions were modeled for all the units.

The results are expressed in internal units: lengths are measured in units of the lattice spacing, σ, and time, τ, is in units of MC steps (MCS) per monomer (on the average each monomer attempts to jump once within a Monte Carlo step, successfully or unsuccessfully). In the simulations, temperature was set to $T = 1$.

2.1. MODEL I: "COLORING" NEAR A PLANAR SUBSTRATE

The homopolymer chain of a fixed length N was placed inside L^3 box ($L_x = L_y = L_z = 240$, in lattice units, σ) with periodic boundary conditions applied in the x and y directions. In the z direction, the system is limited by the two planes at $z = 0$ and $z = L_z$. The $z = 0$ plane is considered as an adsorbing surface, and the $z = L_z$ plane was assumed to be a reflecting wall. The distance between planes is sufficiently large so that the chain cannot interact simultaneously with both planes. To simulate adsorption, we applied a short-range potential, $\varepsilon(z)$, modeling the attraction of monomer units to the surface. The (negative) energy ε (measured in the units of $k_B T$) was attributed to any monomer unit located in the layers $z = 1$ and 2. Thus, to obtain 2δ-height adsorbed layer, we used first two lattice layers. Note that "2δ-adsorption" makes adsorption process more effective and, at the same time, excepts bilayer adsorption. Unadsorbed chain segments were described by a usual three-dimensional self-avoiding walk on a cubic lattice above the surface, while the adsorbed segments were modeled as two-dimensional self-avoiding chains on a square lattice located in the plane $z = 0$. Because the excluded-volume interactions are taken into account for our lattice bond-fluctuation model, each monomer unit occupies one lattice site and prevents other units to invade its nearest environment (i.e., 26 and 8 adjacent sites in $3d$ space and in $2d$ space, respectively). Unoccupied lattice sites are regarded as solvent particles. In this work, the parameter ε was varied from 0 to 6. Note that ε corresponds to the net enthalpy change upon replacement of an adsorbed polymer segment by solvent.

As an initial state, we always prepared a system with zero adsorption potential. When the adsorption potential is switched on, one has to relax the system a long time, since chains need to diffuse from the bulk of the box to the adsorbing wall. Depending on the total number of monomer units in the chain (up to $N = 1024$ in this work), between 10^6 and 10^7 MC time steps per monomer unit were necessary to obtain completely equilibrated configurations. After equilibration at a given segment-surface attraction energy ε, we generated different configurations of a single homopolymer chain containing N segments (all of type A) and interacting with a flat surface with this value of ε. The total number of generated configurations was ranged from 10^7 to 10^8 and each 10^4th of them was recorded by assigning unity to each adsorbed segment ($z \leq 2$) and zero to each unadsorbed segment ($z > 2$). In subsequent studies, these records were treated as copolymer primary structures, where unity indicates a type A segment and zero corresponds to a type B segment. It is evident that at small ε some of the obtained primary structures can consist of only type B segments. The frequency of appearance of these latter primary structures of $N = N_B$ type B segments depends on the parameter of the simulation box L_z. So, in order for the ensemble averages to be independent of such an artificial parameter as the simulation box size, and to reflect the pure effect of the attraction energy at which the ensemble of AB copolymers was generated, we excluded these primary structures from subsequent consideration.

It is clear that the resulting AB composition of the generated copolymers, φ_{AB}, is determined by the average fraction of adsorbed chain segments, φ. One can anticipate that the 1:1 composition will be obtained in the vicinity of the critical adsorption energy of a single homopolymer chain, ε^*. To estimate this energy, we calculated the φ value as a function of ε. The results are shown in Fig. 3.

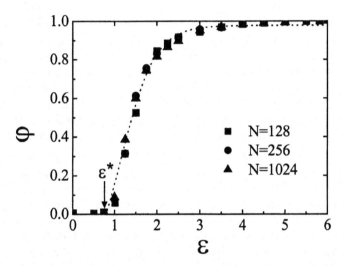

Figure 3. The average fraction of adsorbed (A-type) monomer units against the attraction energy ε (in the units of $k_B T$) for the chains of length $N = 128$, 256, and 1024.

Interception of the ε axis with the straight line tangential to a sigmoidal approximation of the dependence $\varphi(\varepsilon)$ at the inflection point gives an estimate of the critical adsorption energy of the chain of length N [20,21]. It was found that ε^* is practically independent of N for $N \geq 128$ and is estimated as $\varepsilon^* = 0.8\pm0.1$. This value is close to that obtained for the same model of a polymer chain in Refs. [17,21]. However, as seen from Fig. 3, at $\varepsilon = \varepsilon^*$ only ca. 5% of segments are in adsorbed state. From the data presented in Fig. 3 we can conclude that the AB composition approximately corresponding to $\varphi = \frac{1}{2}$ is attained at $\varepsilon \approx 1.5$. Below, we will discuss the adsorption and statistical properties obtained only for these copolymers.

2.2. MODEL II: COPOLYMERIZATION OF SELECTIVELY ADSORBED MONOMERS

In this subsection, we describe a model for radical copolymerization near a surface, which selectively adsorbs polymerizing monomers.

A total of N monomers A and B were placed in a self-closed slab (with periodic boundary conditions in the x and y directions) of size $L_x = L_y = 400$ and $L_z = 120$, in units of σ. The first layer at $z = 1$ was taken as an adsorbing layer and the last layer, $z = L_z$, was considered as a reflecting solid wall. Periodic boundary conditions were applied in the x and y directions. To model adsorption, we used the following adsorption potential: $\varepsilon(z) = -\varepsilon/z^{1/3}$ (where ε is the characteristic adsorption energy at $z = 1$) that acts on the A monomers only. In the model, A monomers had a fixed adsorption energy ($\varepsilon_A > 0$), while B monomers had no attraction to the surface, i.e., $\varepsilon_B = 0$. With this choice, the density of A monomers far from the attractive surface is very small, and thus the precise location of the repulsive wall has a negligible influence on the simulation results. Of course, without this confining repulsive wall a monomer A once released from the attractive surface could escape off toward $z \to \infty$ while with the present geometry it will diffuse in the simulation box until it gets readsorbed at the attractive wall. So the chosen geometry ensures that a stable thermodynamic equilibrium can in principle be achieved.

The simulation was performed in two stages. The first one consisted of modeling free (nonreacting) monomers inside the simulation box and was aimed at calculating the equilibrium adsorption profiles. Among the monomers, N_A were the monomers of type A and N_B were the monomers of type B. We placed into the simulation box a fixed number of A monomers, while the number of B monomers was considered to be a variable parameter. The values of N_A and ε_A were such that about 95% of all A monomers were in strongly adsorbed state for each N_B. The bulk number density of unadsorbed B monomers, $\rho_B = N_B/(L_x \times L_y \times L_z)$, was much larger than the number density of A monomers, $\rho_B \gg \rho_A$. The ρ_B value was varied from 0.01 to 0.25. After long equilibration, the simulations were carried out during 3×10^6 MC steps. After each 10^3 MC steps the monomer density profiles normal to the xy plane were calculated. For the calculation of the density profile, we divided the distance L_z into 120 bins and sampled how many monomers fall into each bin. Then, we calculated the average density profiles for A and B monomers, $\rho_A(z) = N_A(z)/S$ and $\rho_B(z) = N_B(z)/S$, where $S = L_x \times L_y$ is the surface area and $N(z)$ denotes the average number of the corresponding monomers in the z layer. Thus, choosing the number density of B monomers in the bulk, ρ_B, we can obtain various density profiles $\rho_B(z)$. Note that $\rho_A(z)$ depends on ρ_B very weakly. For each ρ_B, nevertheless, the value of ε_A was chosen such that $\rho_A(z=1) = 0.95$ in the first z layer.

During the second phase of the simulation, we studied the irreversible polymerization process, which was considered as a step-by-step chemical reaction of addition of monomer units (A or B) to the growing polymer chain. For the polymerization process, A and B monomers were assumed to be bifunctional. If a free monomer comes within the prescribed distance R_{min} of the end monomer unit of growing chain, a bond can be formed between these two species. Note that, for a radically polymerizing linear polymer, only one of the two ends of the chain is active, and is ready for reacting with free monomers. Also, no polymer rings are allowed to form. In our polymerization model, the presence of low-molecular-weight species (both solvent molecules and free monomers) were taken into account only implicitly through the density profiles of A and B monomers obtained at the first stage. Thus, the model does not contain the information about steady state concentration gradients arising in the course of chain growth. It should be emphasized, however, that for very low monomeric reaction probability, correlations in the density of reactive species are only weakly perturbed by the reactions, and the reaction rate is proportional to the equilibrium contact probability of reactive species. It is assumed that the reactants are ideally mixed and that diffusion is not important. In principle, such conditions can be realized if typical diffusive time, τ_d, and the characteristic time of adsorption equilibrium are essentially less than the reaction time τ_r. It is clear that this assumption corresponds to the kinetically controlled regime. This reaction kinetics is largely independent of the details of the monomeric reaction mechanism, especially of the monomeric capture radius R_{min}.

Usual radical polymerization process consists of the following three parts: initiation, chain propagation, and termination. For our model situation, the initiator of polymerization process was modeled as a fixed point with coordinates $x = L_x/2$, $y = L_y/2$, and $z = 1$ and was fixed on the attractive surface. For simplicity, intermolecular and intramolecular chain transfer reactions were ignored. Since only one radical initiator was put into reaction mixture, spontaneous chain terminating reactions, such as radical combination and disproportionation, were excluded. This situation takes place in reality for the case when the concentration of radical species in a polymerization reaction is very small relative to other reagents (e.g., monomers, solvent molecules, and active chain ends), the rate at which the radical-radical termination reactions occur is also very small, and most growing chains achieve moderate length before termination. In the present study, the growing chain was automatically terminated at the number of monomer units $N = 1024$. Notice that, since in our model the concentration of monomers in solution does not depend on their conversion and, therefore, always remains a constant, we in fact consider an infinitely diluted polymer solution.

The model involves two characteristic time scales. First, the reaction time scale is set by the time between two reaction trials, τ_r. Second, the equilibrium motion time, τ_{eq}, includes conformation relaxation of growing chain. Thus, $\tau_r > \tau_{eq}$ and $\tau_r - \tau_{eq} = \tau_{ad}$, where τ_{ad} is time of monomer addition to the chain. Internal motion of the chain mainly depends on the current chain length, N_{gr}; that is, the length of growing chain corresponding to a given time τ. It should be noted here that the Rouse relaxation time $\tau_{eq} \sim N^2$ for an ideal chain. For simplicity, we take the following fixed relaxation time $\tau_{eq} = 1024 \times N_{gr}$, where N_{gr} is the number of units in growing chain. Radical reactivity and, hence, the value of τ_{ad}, is independent of molecular weight of growing chain. In copolymerization mechanism, the rate of addition of a monomer to a growing free radical depends only on the nature of the group on the chain. There are four possible ways in which a monomer can be added:

Reaction	Rate
$A\bullet + A \rightarrow AA\bullet$	$k_{AA}[A\bullet][A]$
$A\bullet + B \rightarrow AB\bullet$	$k_{AB}[A\bullet][B]$
$B\bullet + A \rightarrow BA\bullet$	$k_{BA}[B\bullet][A]$
$B\bullet + B \rightarrow BB\bullet$	$k_{BB}[B\bullet][B]$

The monomer reactivity ratios r_A and r_B are the ratios of the rate constant for a given radical adding its own monomer to the rate constant for adding the monomer of other type, $r_A = k_{AA}/k_{AB}$ and $r_B = k_{BB}/k_{BA}$. We model an "ideally polymerizing" system, in which two radicals show the same preference of adding one of the monomers over the other, $r_A = r_B$. Moreover, we assume that $k_{AA} = k_{AB} = k_{BA} = k_{BB}$. In this case, the end group on the growing chain has no influence on the rate of monomer addition, and the two types of monomer units are arranged at random along the chain in relative amounts determined by the local concentrations of A and B monomers near the end radical of the chain. Thus, the time of addition of monomer to the chain, τ_{ad}, and probabilities of A or B monomer additions are determined by combination of the equilibrium density profiles of A and B monomers, $\rho_A(z)$ and $\rho_B(z)$, and by the growing chain end location. During the simulation, we found the current position of free end of the chain $[x_p(\tau), y_p(\tau), z_p(\tau)]$ and added to it the new monomer, according to relative probabilities to occur the monomer A or monomer B in the z layer: $p_A = \widetilde{\rho}_A(z_p)$ and $p_B = \widetilde{\rho}_B(z_p)$, where $\widetilde{\rho}_A(z_p)$ and $\widetilde{\rho}_B(z_p)$ are the normalized density profiles.

Thus, in the model used here, we have only one variable parameter of the copolymerization process, viz., the number density of monomers B in the bulk, ρ_B. After termination of the growing chains at $N = 1024$, their statistical analysis was carried out. To gain better statistics, 1000 copolymer chains for each density ρ_B were generated and then required average characteristics were found. Throughout the simulation, snapshots of the system were collected. To give a visual impression of the simulated system, Fig. 4 presents a typical snapshot picture of an 1024-unit copolymer chain synthesized at $\rho_B = 0.16$.

Figure 4. Snapshot of the copolymer chain with $N = 1024$ obtained from copolymerizing monomers at $\rho_B = 0.16$. The white and gray spheres present adsorbed A monomer units and unadsorbed B monomer units, respectively. The place of connection of the chain with the surface ("initiator") and the chain end are depicted as larger spheres.

3. Results and Discussion

3.1. THE STRUCTURE OF ADSORBED COPOLYMERS

Several equilibrium characteristics of the adsorbed copolymer chains were calculated. They include the average numbers of monomer units in train sections, N_{tr}, in loop sections, N_{lp}, and in tail sections, N_{tl}, the average numbers of trains, n_{tr}, loops, n_{lp}, and tails, n_{tl}, in a chain. By a train section of a chain we mean its adsorbed section; a loop is an unadsorbed section between two trains, and a tail is the unadsorbed end section of a chain. Note that $n_{tr} \times N_{tr} + n_{lp} \times N_{lp} + n_{tl} \times N_{tl} = N$. For model I, all these quantities were calculated as a function of the adsorption energy ε, which is gained when a monomer A is in contact with the adsorbing surface. The average numbers of chemically uniform A and B segments in the designed copolymers, m_A and m_B, as well as the average lengths of these segments, L_A and L_B, were found for models I and II. For both copolymers, the length of the segment composed by the units of a given type is a random quantity. First, a thermodynamic average of each of these values was found for a chain with a certain primary structure, and, then, the thermodynamic average was averaged over the sample of the sufficiently large number of primary structures of the corresponding subensemble in which the AB composition φ_{AB} is close to 1:1.

Some of the results obtained for 1024-unit "colored" chains (model I) are shown in Fig. 5.

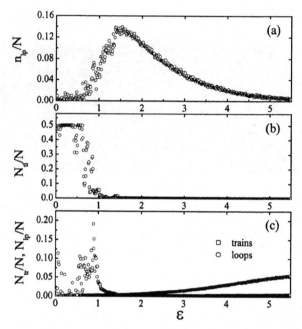

Figure 5. (a) The average number of loops, n_{lp}; (b) the average number of monomer units in tails, N_{tl}, and (c) the average number of monomer units in trains, N_{tr}, and loops, N_{lp}, of an adsorbed 1024-unit copolymer as a function of ε. The results correspond to copolymer chain obtained via the "coloring" procedure (model I).

It is seen that at small values of ε the average number of loops ($n_{lp} = n_{tr}-1$) is close to zero and increases with ε. In the vicinity of ε = 1.5, the $n_{lp}(ε)$ function attains a maximum and then slowly decreases with increasing ε [Fig. 5(a)]. Note that in the strong adsorption regime, nearly all type A segments and some portion of type B segments lie on the surface. So, loops can be formed only from type B segments. It is interesting that the maximum of $n_{lp}(ε)$ is located practically at the same value of ε at which the "coloring" of the bare homopolymer chain was performed. The average tail length quickly decreases with ε and becomes virtually zero in the region ε ≳ 1.5 [Fig. 5(b)]. The average train length N_{tr} is a slowly increasing function of ε. This is quite understandable, because in the strong adsorption regime all type A segments and some portion of type B segments lie on the adsorbing surface. On the other hand, similar to the $n_{lp}(ε)$ function, the average loop length N_{lp} goes through the maximum [Fig. 5(c)]. In this case, however, the maximum is shifted toward smaller ε. Note that the curves $N_n(\varepsilon)$ and $N_p(\varepsilon)$ cross each other at ε ≈ 1.5.

The results presented above allow to assert that the adsorbed copolymer "remembers" the preparation conditions and displays this memory in some adsorption properties discussed here. In addition, as has been mentioned in the Introduction, such a copolymer demonstrates "adsorption-tuned" properties [17].

Now, let us consider some properties of the copolymers obtained via copolymerization near a selectively adsorbing surface (model II).

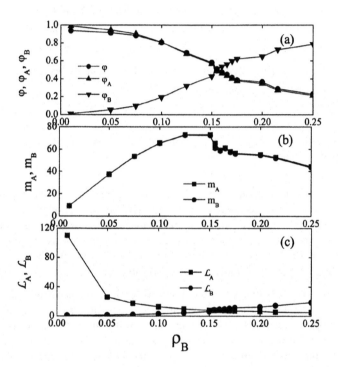

Figure 6. (a) The average fraction of adsorbed monomer units, φ, and the average fractions of A and B units in the designed chain, $\varphi_A = N_A/N$ and $\varphi_A = N_A/N$; (b) the average numbers of A and B blocks, m_A and m_B; and (c) the average lengths of these blocks, $\mathcal{L}_A$ and $\mathcal{L}_B$, as a function of the number density of B monomer in the bulk, ρ_B. Chain length $N = 1024$.

In Fig. 6(a) we plot the average fraction of adsorbed monomer units, φ, as well as the average fractions of A and B units in the designed chain, $\varphi_A = N_A/N$ and $\varphi_A = N_A/N$, as a function of the number density of B monomer in the bulk,

ρ_B. As seen, the curves $\varphi_A(\rho_B)$ and $\varphi_B(\rho_B)$ cross each other at some critical point $\rho_B^* = 0.155$, where we have an AB copolymer with 1:1 composition. Also, it is seen that in this case about half of chain segments is in adsorbed state, $\varphi = \frac{1}{2}$. At $\rho_B < \rho_B^*$, the growing chain is enriched with A monomers; when $\rho_B > \rho_B^*$, the B monomer units prevail in the resulting copolymer (in this case, the growing chain end deeply penetrates into the bulk). Thus, the change in the solution concentration of unadsorbed (or weakly adsorbed) reactive monomers allows varying the composition of synthesized copolymer in a wide range.

The average numbers of A and B blocks, m_A and m_B, and the average lengths of these blocks, $\mathcal{L}_A$ and $\mathcal{L}_B$, are shown in Figs. 6(b) and 6(c) as a function of ρ_B. As seen from Fig. 6(b), the functions $m_A(\rho_B)$ and $m_B(\rho_B)$ are almost identical, and in the region $\rho_B \lesssim \rho_B^*$ they reach a maximum, where $m_A \approx m_B \approx 73$. Next, at $\rho_B \approx \rho_B^*$ we can see [Fig. 6(c)] that the average block lengths become equal to each other, $\mathcal{L}_A \approx \mathcal{L}_B \approx 9$ (in this case, $m_A \approx m_B \approx 57$).

3.2. ANALYSIS OF DESIGNED SEQUENCES: LONG-RANGE CORRELATIONS

Let us now turn to the question of long-range correlations in the copolymer sequences obtained in the present work. In Ref. [10] we have shown that already single "coloring" of a homopolymer globule leads to long-range correlations of the Levy-flight type [11]. The question is: whether these correlations are present in the polymers obtained via the "coloring" of an adsorbed homopolymer chain and in the process of copolymerization near an adsorbing surface?

The statistical analysis of the designed sequences was carried out according to the method similar to that used by Stanley and co-workers [22-24] in their search for long-range correlations in DNA sequences. In this approach, each AB sequence is transformed into a sequence of symbols 0 and 1. Each 0 is interpreted as a downward step and 1 as an upward step of one-dimensional random walk. Shifting the sliding window of length λ step by step along the sequence, the number of A and B monomer units inside the window is counted at each step. This number, which we write as $\gamma(\lambda) = \sum_{i=j}^{j+L} \xi_i$ is a random variable, depending on the position j of the window along the sequence; here ξ_i is the variable associated with every monomer i, such that $\xi_i = 0$ if monomer i is B and $\xi_i = 1$ if it is A. This random variable has certain distribution. Its average is determined by the overall sequence composition (the total numbers of A and B monomers), and its dispersion is given by $D_\lambda^2 = \sum_{i,j=k}^{k+\lambda} \left(\langle \xi_i \xi_j \rangle - \langle \xi_i \rangle \langle \xi_j \rangle \right)$. If the string sequence is uncorrelated (normal random walk) or there are only local correlations extending up to a characteristic range (Markov chain) then the value of D_λ scales as $\lambda^{1/2}$ with the window of sufficiently large λ. A power law $D_\lambda \propto \lambda^\alpha$ with $\alpha > 1/2$ would then manifest the existence of long-range (scale-invariant) correlations. However, due to large fluctuations, conventional scaling analyses of the dispersion D_λ cannot be applied reliably to the entire short sequence. To avoid this problem, we use the detrended fluctuation analysis (DFA), the method specifically adapted to handle problems associated with nonstationary sequences [23,24]. In brief, DFA involves the following steps: (i) Divide the entire N-symbol string into N/λ nonoverlapping substrings, each consisting λ symbols, and then define the local trend in each substring to be the ordinate of a linear least-squares fit for the random walk displacement within this substring. (ii) Find the difference $\Delta(\lambda)$ between the original random walk $\gamma(\lambda)$ and the corresponding local trend. (iii) Calculate the variance about the detrended walk for each substring. The average of these variances over all the substrings of size λ, denoted $F_D^2(\lambda)$, characterizes the detrended local fluctuations within the sliding window of length λ. Generally, the $F_D(\lambda)$ function shows the same behavior as D_λ.

Figure 7 presents the results of the statistical analysis performed as described above for 1000 sequences obtained via the "coloring" procedure. On average, these sequences have 1:1 AB composition and block lengths $\mathcal{L}_A \approx \mathcal{L}_B = 4.4$.

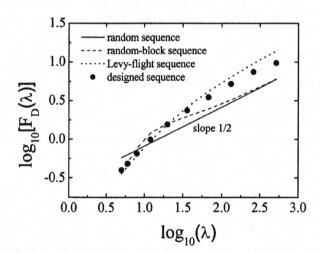

Figure 7. Detrended fluctuation function for random, random-block, Levy-flight, and designed (model I) sequences of length $N = 1024$.

For comparison, we show in the same figure the data obtained for a purely random 1:1 sequence and a random-block 1:1 sequence, both with $N = 1024$. We note that by a random sequence we mean that for which the probability p_A of each unit to be of type A is constant throughout the whole sequence. This means, for example, that the conditional probability $p_A(i+1,i)$ that the $(i+1)^{th}$ unit is of type A when the ith unit is also of type A is equal to p_A. The average fraction of type A units in a random sequence is equal to the probability p_A, which in our case is set to $p_A = ½$. It is clear that for such sequence $L_A = L_B = 2$. By definition, a random-block sequence is characterized by the Poisson distribution of the type A and B block lengths, $f(x) = e^{-L} L^x / x!$, where L is the average block length (L_A for type A blocks and L_B for type B blocks). Systematic consideration and comparison of the properties of the ensembles implies that we must determine L_A and L_B for the ensemble of designed sequences, and create the random-block ensemble using these parameters in the corresponding Poisson distribution. Thus, we generated the Poisson distribution adjusted to achieve the same 1:1 composition and the same "degree of blockiness" as for a designed copolymer obtained by "coloring". Also, we present in Fig. 7 the $F_D(\lambda)$ function found for a model Levy-flight sequence [10] of length $N = 1024$.

It is seen from Fig. 7 that the data obtained for a purely random sequence demonstrates $F_D(\lambda) \propto \lambda^{½}$ scaling throughout the interval of λ examined, as expected. For a random-block sequence having the same composition and average block length as designed sequences, there are well-pronounced correlations in the region corresponding to small λ's, which are close to the average block length. However, as λ increases, we observe dumping of the correlations, and for sufficiently large λ the behavior demonstrated by the random and random-block sequences is almost identical. Comparing these curves with the simulation results we see immediately that designed sequences do not correspond to random and random-block statistics, and strong correlations do exist in these sequences. Although the simulation results do not fit accurately to any power law $F_D(\lambda) \propto \lambda^{\alpha}$ throughout the interval of λ considered, the slope α observed for the dependence of $\log[F_D(\lambda)]$ on $\log\lambda$ corresponds to a value significantly larger than $½$, up to about 1 at small values of λ, thus indicating pronounced long-range correlations in these designed sequences. Of course, when λ becomes comparable to the total chain length N, the long-range correlations fade and $\alpha \to ½$. A similar behavior with a wide crossover region between $\alpha \to 1$ and $\alpha \to ½$ is seen for a model sequence with the Levy-flight statistics (Fig. 7). Thus, we conclude that the "coloring" of adsorbed chains leads to the sequences, which distinctly demonstrate the presence of long-range correlations and these correlations are of the Levy-flight type. In this respect the properties of the copolymers obtained by "coloring" of partially adsorbed homopolymer chains are similar to those known for proteinlike copolymers [10].

In Fig. 8 we compare the statistical behavior of sequences obtained by co-polymerization (model II) with that demonstrated by several model sequences. As has been noted in the previous subsection, under conditions considered in this study, the average block lengths in the designed sequences are $L_A \approx L_B \approx 9$ when the 1:1 composition is realized for an ensemble of 1024-unit copolymers. Thus, the same block length was used to generate an ensemble of model random-block sequences.

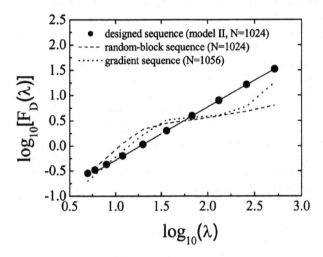

Figure 8. Detrended fluctuation function for designed (model II), random-block, and gradient sequences.

For the designed sequences, we find pronounced long-range correlations, which persist up to the windows with length close to N. In contrast to the sequences obtained via "coloring" procedure (model I), these sequences do not show a wide crossover region with varying slope: as seen, the dependence of $\log[F_D(\lambda)]$ on $\log\lambda$ is a nearly linear function (with the slope close to 1) almost for all the λ's considered. Therefore, the statistics of synthesized copolymers differs from the Levy-flight statistics [11]. Of course, it also differs from random and random-block distributions. Thus, the question is: what kind of statistics corresponds to our designed copolymers?

To gain some insight into the problem discussed, we have monitored the variation of AB composition during copolymerization. It was found that the AB composition changes along the growing chain during copolymerization. At the beginning of copolymerization, the growing chain radical gathers mainly A monomers strongly adsorbed at the surface. Then, as the chain becomes longer and the number of connected B units increases, the probability to find growing end-radical in B environment rises, resulting in an increase in the frequency of reactions with the participation of unadsorbed B monomers the concentration of which dominates in the bulk. This causes the changes both in the current AB composition and in the average block lengths in growing chains of a given length N_{gr}. Figure 9 illustrates these trends. Here, we present the current average block lengths, $L_A(N_{gr})$ and $L_B(N_{gr})$, vs. the average length of growing copolymer fragment, N_{gr}. As seen, the $L_A(N_{gr})$ value reaches its limiting level L_A rather quickly (at $N_{gr} > 15$). On the other hand, the current average length of blocks B monotonically increases with N_{gr}. As a result, for our copolymer, which was adjusted to achieve 1:1 composition and required length of $N = 1024$, we obtain the primary structure of gradient-like type. In this copolymer, the content of B monomer units and the corresponding block length gradually increase from the beginning of chain to its end, while the same quantities observed for A monomer units remain practically unchanged along the chain. It is clear that such a situation can take place only if both the total chain length and AB composition are constrained simultaneously. Without these constraints we would have: $L_B \rightarrow \infty$, $L_A \rightarrow constant$, and $\varphi_A \rightarrow 0$ for the $N \rightarrow \infty$ limit. It should be emphasized that the statistical properties of synthesized copolymers strongly deviate from that observed for an ideal gradient sequence with 1:1 composition [25]. This fact is illustrated in Fig. 8, where we show the detrended fluctuation function calculated for such sequence.

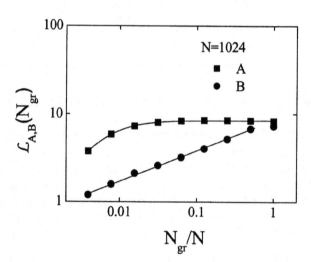

Figure 9. Current average block lengths, $L_A(N_{gr})$ and $L_B(N_{gr})$, vs. the average length of growing copolymer fragment, N_{gr}, for 1024-unit chains (model II).

4. Conclusion

In this work, we have used a Monte-Carlo simulation technique to study two-letter (AB) quasirandom copolymers with quenched primary structure, which were generated via surface-induced computer-aided sequence design, following a general concept first introduced by us for a solution of polymer globules [1-4] (for a recent review, see Ref. [26]). Two different approaches have been employed: (i) a "coloring" (i.e., chemical modification) of partially adsorbed homopolymer chains and (ii) a copolymerization near a surface. In the former case, AB copolymers were prepared by adsorbing a bare homopolymer chain onto a flat substrate, after which its adsorbed segments were converted to type A ones, and its unadsorbed segments to type B ones. The second approach represents an irreversible radical copolymerization of selectively adsorbed A and B monomers with different affinity to a surface. We have shown that the statistical properties of the copolymer sequences obtained in these two ways exhibit well-pronounced long-range correlations in distribution of different monomer units along the chain. In particular, the primary structure of copolymers generated via the "coloring" procedure is similar to that known for proteinlike copolymers [1-4,10]. On the other hand, the copolymers obtained from polymerizing monomers selectively adsorbed onto a flat substrate demonstrate a specific quasi-gradient primary structure under the preparation conditions studied in this work. In addition, some characteristics of the adsorbed single chains (statistics of trains, loops, and tails) have been also investigated.

Acknowledgements

We would like to thank S.I. Kuchanov and A.V. Berezkin for stimulating discussions. The financial support from Alexander-von-Humboldt Foundation, Program for Investment in the Future (ZIP), INTAS (project # 01-607), SFB 569, and Russian Foundation for Basic Research is highly appreciated.

References

1. Khokhlov, A.R. and Khalatur, P.G. (1998) Proteinlike copolymers: Computer simulation, *Physica A* **249**, 253-261.
2. Khalatur, P.G., Ivanov, V.I., Shusharina, N.P., and Khokhlov, A.R. (1998) "Proteinlike" copolymers: Computer simulations, *Russ. Chem. Bull.* **47**, 855-860.
3. Khokhlov, A.R., Ivanov, V.A., Shusharina, N.P., and Khalatur, P.G. (1998) Engineering of synthetic copolymers: Proteinlike copolymers, in: F. Yonezawa, K. Tsuji, K. Kaij, M. Doi, and T. Fujiwara, (eds.), *The Physics of Complex Liquids*, World Scientific, Singapore, p. 155.
4. Khokhlov, A.R. and Khalatur, P.G. (1999) Conformation-dependent sequence design (engineering) of AB copolymers, *Phys. Rev. Lett.* **82**, 3456-3460.

134

5. Lifshitz, I.M., Grosberg, A.Yu., and Khokhlov, A.R. (1978) Some problems of the statistical physics of polymer chains with volume interactions, *Rev. Mod. Phys.* **50**, 683-713.
6. Grosberg, A.Yu. and Khokhlov, A.R. (1994) *Statistical Physics of Macromolecules*, American Institute of Physics, New York.
7. Khalatur, P.G., Khokhlov, A.R., Mologin, D.A., and Reineker, P. (2003) Aggregation and counterion condensation in solution of charged proteinlike copolymers: A molecular dynamics study, *J. Chem. Phys.* **119**, 1232-1247.
8. Zherenkova, L.V., Khalatur, P.G., and Khokhlov, A.R. (2003) Solution properties of charged quasi-random copolymers: Integral equation theory, *J. Chem. Phys.* **119**, 6959-6972.
9. Ivanov, V.A., Chertovich, A.V., Lazutin, A.A., Shusharina, N.P., Khalatur, P.G., and Khokhlov, A.R. (1999) Computer simulation of globules with microstructure. *Macromolecular Symposia* **146**, 259-265.
10. Govorun, E.N., Ivanov, V.A., Khokhlov, A.R., Khalatur, P.G., Borovinsky, A.L., and Grosberg, A.Yu. (2001) Primary sequences of proteinlike copolymers: Levy-flight-type long-range correlations. *Phys Rev E* **64**, 0409031(R)-4(R).
11. Shlesinger, M.F., Zaslavskii, G.M., and Frisch, U. (1996) Levy flights and related topics in physics. *Lecture Notes in Physics*, Springer-Verlag, Berlin.
12. Virtanen, J., Baron, C., and Tenhu, H. (2000) Grafting of poly(N-isopropyl acrylamide) with poly(ethylene oxide) under various reaction conditions, *Macromolecules* **33**, 336-431.
13. Virtanen, J., Lemmetyinen, H., and Tenhu, H. (2001) Fluorescence and EPR studies on the collapse of poly(N-isopropyl acrylamide)-*g*-poly(ethylene oxide) in water, *Polymer* **42**, 9487-9493.
14. Lozinsky, V.I., Simenel, I.A., Kurskaya, E.A., Kulakova, V.K., Grinberg, V.Ya., Dubovik, A.S., Galaev, I.Yu., Mattiasson, B., and Khokhlov, A.R. (2000) Synthesis and properties of "proteinlike" copolymers, *Dokl. Chem.* **375**, 273-276.
15. Wahlund, P.-O., Galaev, I.Yu., Kazakov, S.A., Lozinsky, V.I., and Mattiasson, B. (2002) "Proteinlike" copolymers: Effect of polymer architecture on performance in bioseparation process, *Macromol. Biosci.* **2**, 33-42.
16. Berezkin, A.V., Khalatur, P.G., and Khokhlov, A.R. (2003) Computer modeling of synthesis of proteinlike copolymer via copolymerization with simultaneous globule formation, *J. Chem. Phys.* **118**, 8049-8060.
17. Zheligovskaya, E.A., Khalatur, P.G., and Khokhlov, A.R. (1999) Properties of AB copolymers with a special adsorption-tuned primary structure, *Phys. Rev. E* **59**, 3071-3078.
18. Carmesin, I. and Kremer, K. (1988) The bond fluctuation method: A new effective algorithm for the dynamics of polymers in all spatial dimensions, *Macromolecules* **21**, 2819-2823.
19. Allen, M.P. and Tildesley, D.J. (1990) *Computer Simulation of Liquids*, Claredon Press, Oxford.
20. Kramarenko, E.Yu., Winkler, R.G., Khalatur, P.G., Khokhlov, A.R., and Reineker, P. (1996) Molecular dynamics simulation study of adsorption of polymer chains with variable degree of rigidity: Static properties, *J. Chem. Phys.* **104**, 4806-4813.
21. Zheligovskaya, E.A., Khalatur, P.G., and Khokhlov, A.R. (1997) Polymer chain binding with a flat adsorbent in the case of selective adsorption of segments: Monte Carlo simulation, *J. Chem. Phys.* **106**, 8598-8605.
22. Peng, C.K., Buldyrev, S.V., Goldberger, A.L., Havlin, S., Sciortino, F., Simon, M., and Stanley, H.E. (1992) Long-range correlations in nucleotide sequences, *Nature* (London) **356**, 168-171.
23. Peng, C.K., Buldyrev, S.V., Goldberger, A.L., Havlin, S., Simons, M. and Stanley, H.E. (1993) Finite size effects on long-range correlations: Implications for analyzing DNA sequences, *Phys. Rev. E* **47**, 3730-3733.
24. Buldyrev, S.V., Goldberger, A.L., Havlin, S., Peng, C.-K., Simons, M., and Stanley, H.E. (1993) Generalized Levy walk model for DNA nucleotide sequences, *Phys. Rev. E* **47**, 4514-4523.
25. Davis, K.A. and Matyjaszewski, K. (2002) *Statistical, Gradient, Block and Garft Copolymers by Controlled/Living Radical Polymerizations*, Springer, Berlin and New-York.
26. Khokhlov, A.R. and Khalatur, P.G. (2003) Biomimetic sequence design in functional copolymers, *Curr. Opin. Solid State & Materials Sci.*, in press.

Novel approach to the study of rotational and translational diffusion in crystals

A.V.BELUSHKIN
Frank Laboratory of Neutron Physics
JINR
141980 Dubna
Russia

1. Introduction

Among modern research techniques for the study of structure of crystals most widely used are X-ray diffraction and synchrotron radiation, as well as neutron diffraction which often allows supplementing the data of first two methods. From the practical point of view the diffraction on powder samples will more and more widely be used, as the new materials are frequently synthesised in small quantities not permitting to grow a single crystal. And in a series of cases the growth of single crystals encounters serious difficultiesfor other reasons. Therefore methods of the analysis of powder diffractograms have received major development, including the *ab-initio* structure solution of new complex crystalline compounds (see e.g. [1]), and very precise refinement of structure, primarily determined from the analysis of a diffraction on single crystals. As an example, let us briefly discuss the profile refinement procedure for powder diffractograms based on so called Rietveld method [2]. Below the discussion will be limited to neutron diffraction technique, therefore all formulas and arguments are related to this type of experiments.

At the profile analysis of the diffraction data obtained on powder samples, the experimental histogram is generally modelled by the following parameterised function (see e.g. [3]):

$$I_i^c = I_i^b + \sum_{m=1}^{Nb} S_m \sum_{k=k_1}^{k_2} j_{mk} \, Lp_{mk} \, O_{mk} |F_{mk}|^2 \Omega_{imk} \qquad (1)$$

Where I_i^b - intensity of a background for the *i-th* point of a histogram
S_m - volume fraction of a phase of type m (for the multiphase refinement)
j_k – multiplicity factor of the *k-th* reflection, that is the number of combinations of

Miller indices (hkl), corresponding to identical interplanar distances.
Lp_k – Lorentz factor describing a particular geometry and type of diffraction experiment
O_k – function describing a texture that is the deviation from purely random distribution of relative grain orientation in the sample.

A.T. Skjeltorp and A.V. Belushkin (eds.), Forces, Growth and Form in Soft Condensed Matter: At the Interface between Physics and Biology, 135-144.
© 2004 *Kluwer Academic Publishers. Printed in the Netherlands.*

distribution of relative grain orientation in the sample.

$|F_k|$ - module of a structure factor corrected for the thermal vibrations of atoms.

Ω_{ik} - function featuring the diffraction peak profile, including the resolution of the instrument.

Further procedure of data processing is reduced to minimisation of a functional

$$\chi^2 = \sum_i \frac{1}{\sigma_i} (I_i^{exp} - I_i^c)^2,$$

(2)

Where I_i^{exp} is the experimental value and σ_i is the variance i-th point of a measured histogram and the total sum is over all experimental data points.

As a result of data processing the structural model of a crystal is obtained, which corresponds (on the basis of certain criteria) most adequately to the experimental data. As it was already mentioned above, modern methods of the analysis of powder diffractograms, including the method of the profile analysis briefly outlined above, allow one to study rather complex crystal structures. However alongside with a lot of problems of methodical and mathematical character, which one in the given paper we shall not concern, the conventional methods of data analysis leave behind a problem of disorder in crystals, which can have both static and dynamic character. As an example, fig.1 shows a neutron diffraction spectrum obtained from CsDSO$_4$ sample in its high proton conductivity phase. Modulated diffuse scattering arising due to deuteron translational and SO$_4$ rotational diffusion is clearly seen.

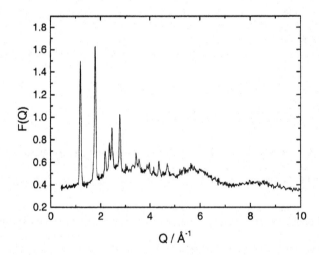

Figure 1. Neutron diffraction spectrum obtained from CsDSO$_4$ sample in its high proton conductivity phase. The modulated diffuse scattering under the Bragg peaks is clearly seen.

The problem is connected with the fact that the effects of disorder contribute as a diffuse scattering to the intensity of background I_i^b in formula (1). Usually background is approximated by a polynomial or some other smooth function in the case of conventional data analysis and is not treated any further. To extract the physical information from the diffuse scattering it is necessary to use alternative methods. One of the approaches to solve the problem is the inverse Monte Carlo method that is actively developing at present.

2. Reverse Monte Carlo method

In reverse Monte Carlo (RMC) method [4] the structure of the object under study is described with the help of pair radial correlation function, defined as follows:

$$g(r) = \frac{n(r)dr}{4\pi n_0 r^2 dr}$$

(3)

Here n_0 - average atomic density of the object, $n(r)$ - number of atoms in a spherical layer of radius r and thickness dr, constructed around of atom, arbitrary selected as an origin.

In diffraction experiment Fourier transform of correlation function $g(r)$ is measured (after necessary calibrations and corrections):

$$S^{exp}(Q) = \frac{4\pi n_0}{Q} \int_0^{+\infty} r\, g(r)\, sin(Qr)dr.$$

(4)

$S^{exp}(Q)$ is called the total structure factor, Q - a neutron momentum transfer.

Avoiding specific details, the procedure of data analysis in reverse Monte Carlo method is the following. At the first stage the structural model is constructed, which one, as a rule, is based on the previously known data or justified assumptions about the object. For this model the pair radial correlation function and relevant total structure factor $S_0^{calc}(Q)$ are calculated. Then, calculated and experimental structure factors are compared on the basis of maximum likelihood criteria:

$$\chi_0^2 = \sum_i \{ S_0^{calc}(Q_i) - S^{exp}(Q_i) \}^2 / \sigma^2(Q_i),$$

Where the sum is over all points of experimental spectrum included in a data analysis. $\sigma(Q_i)$ –the variance of i-th point of an experimental data. On a following step one atom is selected at random and displaced on arbitrary quantity in an arbitrary direction. For the structural model obtained the new correlation function $g_n(r)$, total structure factor $S_n^{calc}(Q)$ and relevant parameter χ_n^2 are calculated. In case when χ_n^2 is smaller than for the previous configuration, the new structural model is accepted. In the opposite case, the new model is accepted with probability $w = exp \left[-(\chi_n^2 - \chi_{n-1}^2)/2\right]$ and is rejected otherwise.

As a result, parameter χ_n^2 decreases to some minimum quantity, which one corresponds to structural model that explains most adequately experimental data. This algorithm implements so-called "Markovian process", therefore final structural model appears to be statistically independent from initially selected starting three-dimensional configuration of atoms.

Among the relevant advantages of reverse Monte Carlo method we shall mark an opportunity to analyse structure of strongly disordered objects, including fluids and glasses that is impossible with the help of the conventional profile analysis described in the previous section. Another advantage is the absence of necessity of any assumptions about character of interatomic potential (in contrast, for example, to a method of molecular dynamics).

However, as well as any other method of the analysis of experimental data, the reverse Monte Carlo method is not free and from deficiencies. Those who would like to learn more about opportunities and difficulties of application of the method can refer to paper [4] and other publications in the same issue.

3. Application of reverse Monte Carlo method for the analysis of a rotational and translational diffusion in crystals

3.1 AMMONIUM CHLORIDE NH₄Cl

Ammonium chloride NH_4Cl (ND_4Cl) is a good model object for theoretical investigations of the order disorder type phase transitions in molecular -ionic and molecular crystals. Ammonium chloride is a typical system with orientational disorder. At room temperature it has a CsCl-type cubic structure, space group *Pm3m*, in which ammonium ions are randomly distributed between two equivalent positions which correspond to two possible orientations of ammonium tetrahedra. Hydrogen (deuterium) atoms are directed essentially towards four of the eight surrounding halide atoms. Thermal fluctuations of the lattice excite vibrations of ammonium ions around the equilibrium orientations – i.e. librational motion, and also reorientations between the two allowed orientations. The mechanism of reorientation process from the crystal symmetry point of view can involve either 90° jumps around the two-fold axis of NH_4 (ND_4) ions or 120° jumps around the three-fold axis of the tetrahedron or combination of these processes. To resolve the question about the geometry of ammonium ion reorientations and obtain more detailed information about disorder in the high temperature phase of ammonium chloride the reverse Monte Carlo method was applied.

The crystal structure obtained in earlier diffraction studies [5] was used as a starting model. The RMC analysis allowed obtaining the space average deuterium density in the unit cell, which has a form of a cube [6] (fig.2). The corners of the cube correspond to the average crystallographic deuterium positions and are broadened by librational motion of the ammonium ions. At approximately ¼ of the maximum D density these corners then connect, approximately along the edges of the cube. This is the path expected for jumps around two-fold axes. The path expected for three-fold jumps passes quite close to the face centre of the cube where RMC analysis shows much lower D density. A rough estimate allows concluding that the probability for three-fold reorientations is less than 5%.

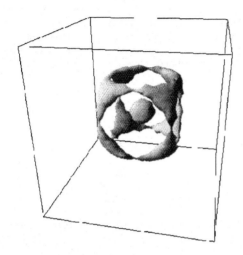

Figure.2. Space average D density distribution in the ND$_4$Cl unit cell.

The results obtained clearly illustrate that the orientational disorder of ammonium ions mimic the cubic symmetry of surrounding halide atoms and the jumps around two-fold symmetry axes of ammonium ions dominate. This is quite consistent with what is expected from simple considerations of the ionic potential [7] and also with quasielastic neutron scattering studies of NH$_4$Br [8] which is isostructural with NH$_4$Cl in a dynamically disordered phase.

3.2. AMMONIUM IODIDE NH$_4$I.

Ammonium iodide NH$_4$I (ND$_4$I) represents more complex case of orientation disorder, than ammonium chloride. Two cubic phases showing orientational disorder of ammonium ions are known to exist in this compound: NaCl-type phase I and CsCl-type phase II. The high temperature NaCl-type phase is highly disordered and the information about its structure is very limited. In this structure (space group $Fm\overline{3}m$), tetrahedral NH$_4$ (ND$_4$) ions are located in the octahedral environment of the surrounding halide ions. In this phase it is not possible to arrange all hydrogen (deuterium) atoms at equal distances to halide atoms, as it was the case for ammonium chloride. Several models were proposed to describe the structure of the high temperature phase of ammonium iodide, including free rotation of ammonium ions, uni-axial diffusion, three-fold uni-axial rotation and several octahedral jump models. However, none of these descriptions could be considered as an appropriate solution.

To obtain reliable information it is necessary to apply methods sensitive to the instantaneous crystal structure, i.e. methods allowing studying deviations from the average crystal structure. And RMC method is very well suited for this purpose.

Neutron diffraction data obtained from ND_4I were first analysed using the standard profile analysis method and the best structural model obtained was used a starting structure. In this model, which was first considered in the structure study of mixed crystals $(NH_4I)_{0.73}(KI)_{0.27}$ [9], three out of four N-D bonds of ammonium ions are oriented (with slight inclinations) towards the nearest halide atoms and the fourth almost coincides with the spatial diagonal of cubic crystal unit cell (fig.3). The spatial average deuterium density distribution obtained from RMC appeared to have the shape of an octahedron, thus mimicking the symmetry of nearest neighbours environment formed by iodine atoms (fig.4).

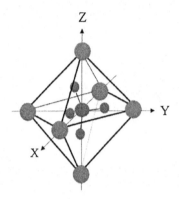

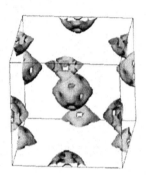

Figure 3 Octahedral environment of the ammonium ion in ND_4I structure. One possible orientation of ND_4 ion is shown .

Figure 4. Space average D density distribution in the the ND_4I unit cell,

Due to librations and reorientations of ammonium ions, the resulting density distribution consists of wide spherical deuterium distributions centred along <111> directions which are connected with the nearest square-shaped D distributions centred along <100> directions. Such a picture suggests that ammonium ions reorient by 39° jumps of N-D bonds between the nearest orientations. In quasielastic neutron scattering measurements of the phase I of NH_4I [10] it was found that the data could be modelled correctly only if one assumes ammonium ions reorientations over angles smaller than 90° but the exact value was not possible to obtain.

The two examples discussed above clearly illustrate the fact that in the orientationally disordered phases system tends to avoid a contradiction between local symmetry of a molecule and global symmetry of its environment. It becomes possible due to the dynamical disordering of the molecule orientation in space so that the atomic density matches fully the crystal symmetry.

3.3 SUPERIONIC CONDUCTIVITY IN CsHSO$_4$.

Caesium hydrogen sulphate CsHSO$_4$ and its deuterated form CsDSO$_4$ are the most extensively studied crystals among the family of superprotonic conductors with hydrogen bonds. It was established that deuteration practically has no effect on the superprotonic phase transition temperature and microscopic mechanism of phase transition (for a review see [11]). CsHSO$_4$ undergoes a first order phase transition into the superprotonic phase at T_c = 414 K and for the deuterated form at T_c = 412 K. In the phase transition the conductivity of crystals increases by four orders of magnitude and reaches $10^{-2}\Omega^{-1}$ cm^{-1} [12]. The phase transition is also characterised by a large spontaneous shear strain (10^{-2}) [13] that complicates the use of single crystals for experimental studies.

The structural mechanism of the superprotonic phase transition was studied by high-resolution neutron powder diffraction using a deuterated sample [14]. In the low conductivity phase the crystal is monoclinic, space group P2$_1$/c and the SO$_4$ tetrahedra are linked by hydrogen bonds so that zigzag chains propagate along the [001] crystallographic direction. The D atoms and SO$_4$ tetrahedra are fully ordered in this phase. The hydrogen bond is strongly non-linear with an O-D-O angle of 166.5° and D atom is positioned 0.68 Å nearer to one O atom than to the other. When the temperature increases above the phase transition the crystal becomes tetragonal, space group I4$_1$/amd. In this phase each SO$_4$ tetrahedron can have four different orientations with equal probability and, therefore, is disordered over them. As a result, the number of crystallographically equivalent positions for each deuteron increases from one in the low conductivity phase to six in the high conductivity phase. The hydrogen bond network becomes disordered and unoccupied proton positions provide the possibility for proton hopping through the lattice.

The diffraction process of protons in the superprotonic phase of CsHSO$_4$ was investigated by incoherent quasi-elastic neutron scattering [15]. On the basis of analysis of the data, taking into account the structural peculiarities of the superprotonic phase, the following two-stage mechanism of proton transport was proposed. The first stage consists in displacing a proton along the hydrogen bond closer to one of two SO$_4$-groups participating in this bond. This results in the formation of a HSO$_4^-$ defect. The second stage consists in breaking the longer half of the hydrogen bond and reorienting the HSO$_4^-$ group to a new position. This reorientation process continues until the second SO$_4$ group is found in the orientation appropriate for the formation of a new hydrogen bond. Then the process is repeated and, as a result, long-range proton transport through the crystal lattice is realised. Schematically this is shown in fig.5, where one of the possible proton diffusion routes is shown. The proposed model of proton diffusion and parameters of this process are in very good agreement with the available experimental data obtained by different methods: calorimetric, optical and neutron spectroscopy and NMR. Nevertheless, since the profile analysis of the diffraction data can give only the average structure of the sample and the interpretation of the quasi-elastic data is based on a certain model, it seems necessary to have an independent, model free verification of the predictions made. In particular, the verification of the fact that the proton position in hydrogen bonds has two minima was beyond the limit of precision of the high-resolution powder neutron diffraction data. The question of the static or dynamic nature of the SO$_4$ disorder in the superprotonic phase also deserves independent study. The last but not the least is the problem of direct observation of the distribution of proton density in the superprotonic phase to identify the proton diffusion channels in the lattice. The reverse Monte Carlo (RMC) modelling of elastic neutron scattering data offers the solution to all these problems.

142

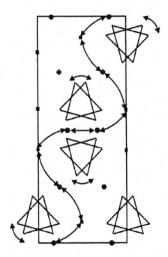

Figure.5. Schematic representation of the proposed mechanism for proton conduction in the high temperature phase of CsDSO$_4$. Two of the four possible SO$_4$ orientations and one of many possible D diffusion pathways are shown. D positions are indicated by filled circles and squares

From the deuteron density map obtained as a result of RMC [16] (fig.6) analysis one can clearly see the percolation paths which ensure proton diffusion through the lattice. If we decrease the probability cut-off limit for deuterons than their distribution becomes looser and occupies larger volume in the unit cell. But in any case this distribution does not overlap with the distribution of other atoms. This fact is an extra proof of the validity of RMC results.

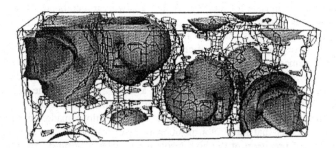

Figure.6. Three dimensional plot of oxygen (closed surfaces) and deuterium (a net) atomic densities in the unit cell for the high conductivity phase of CsDSO$_4$.

The density map for oxygen atoms in the high conductivity phase clearly shows that density forms a closed surface proving the reorientation of SO_4 tetrahedra predicted by structural and quasielastic neutron scattering data. The density isosurface is not spherical and the observed humps correspond very well to the most probable location of oxygen atoms as found from profile refinement of diffraction data.

In conclusion of this section one can say that complex use of different experimental techniques allowed establishing detailed microscopic mechanism of superprotonic phase transition and the mechanism of proton transport. Reverse Monte Carlo method proved to be very efficient complementary tool in this study.

4. Conclusions

The complex use of standard methods (for example, Rietveld structure refinement) and non-conventional method of structure analysis, such as reverse Monte Carlo, provides unique capabilities for investigation of complex crystal structures with dynamic disorder of some molecular groups or atoms. The standard methods of a structure analysis allow constructing structural model of ordered crystalline "skeleton" and to spot equilibrium (most probable) positions of disordered molecular groups and atoms. Then with the help of reverse Monte Carlo method it is possible to visualise and to investigate geometry and parameters of disorder. It is then possible to compare them with the results obtained by other methods. It seems obvious that the potential of the method to explore kinetics of chemical reactions in solutions and solid state, to study conformation changes of molecules and for many other fields of research is enormous and will be realised very soon.

5. Acknowledgements

Author expresses sincere thanks to Doctors R.L.McGreevy, P.Zetterstrom and D.P.Kozlenko for fruitful co-operation in performing the research described in the paper. The support of the Rutherford - Appleton Laboratory (Great Britain) and neutron laboratory NFL Studsvik (Sweden) staff during the experiments as well as provision of access to the facilities by the Directorates of these Laboratories is greatly acknowledged.

6. References

[1]   Wessels T. et.al. J.Am.Chem.Soc. 121 6242 (1999)

[2]   Rietveld H. J. Appl. Cryst. 2 65 (1969)

[3]   McCusker L.B., Dreele R.B., Cox D.E., Louer D., Scardi P. J.Appl.Cryst. 32 36 1999)

[4]   McGreevy R.L. Nucl. Instr. Meth. Phys. Res. 354 1 (1995).

[5]   Yelon W.B., Cox D.E., Kortman P.J., Daniels W.B. Phys. Rev. B 9 4843 (1974).

[6]   Belushkin A.V., Kozlenko D.P. McGreevy R.L, Savenko B.N., Zetterstrom P. Physica B 269 297 (1999).

[7]   Gutowsky H.S., Pake G.E., Bersohn R. J. Chem. Phys. 22 643 (1954).

[8]   Lechner R.E., Badurek G., Dianoux A.J., Herve H., Volino F. J. Chem. Phys. 73 934 (1980).

[9]   Paasch M., McIntyre G.J., Reehuis M., Sonntag R., Loidl A. Z. Phys. B 99 339 (1996).

[10]   Goyal P.S., Dasannacharya B.A. J. Phys. C: Solid State Phys. 12 219 (1979).

[11]   Belushkin A.V. Crystallography Reports 42 501 (1997).

[12]   Baranov A.I., Shuvalov L.A., Shchagina N.M. JETP Lett. 36 459 (1982).

[13]   Yokota S. J. Phys. Soc. Japan 51 1884 (1982).

[14]   Belushkin A.V., David W.I.F., Ibberson R.V., Shuvalov L.A. Acta Cryst. B47 161 (1991).

[15]   Belushkin A.V., Carlile C.J., Shuvalov L.A. J. Phys.: Condens. Matter 4 389 (1992).

[16]   Zetterstrom P., Belushkin A.V., McGreevy R.L., Shuvalov L.A. Solid State Ionics 116 321 (1999).

The Bacterial Flagellar Motor

RICHARD M. BERRY
Department of Physics
University of Oxford
The Clarendon Laboratory,
Parks Road,
Oxford OX1 3PU
UK

1. Introduction

Many species of bacteria actively navigate their environment by swimming (Macnab, 1996, Blair, 1995). Many different swimming styles exist, but most are based on the rotation of bacterial flagella. In all cases flagella are passive, essentially rigid helices, propelled by a rotary motor embedded in the cell envelope at their base (Berry and Armitage, 1999, Macnab, 1996, Blair, 1995). The bacterial flagellar motor is powered by the flow of ions across the inner, or cytoplasmic, membrane of a bacterial cell envelope. Ion flux is driven by an electrochemical gradient, the protonmotive force (pmf) or sodium-motive force (smf) in motors driven by H^+ and Na^+ respectively. The electrochemical gradient consists of a voltage component and a concentration component, and is a key intermediate in the metabolism of both bacteria and higher organisms (the inside of a bacterial cell is typically at an electrical potential about 150 mV below the outside, and also typically has a slightly lower concentration of H^+ or Na^+ ions). Filaments rotate at speeds up to and beyond 1000 Hz in swimming cells and each motor has a maximum power output on the order of 10^{-15} W, two or three orders of magnitude higher than other known molecular motors which are driven by ATP hydrolysis. If cells are attached to a surface by a single flagellar filament, or "tethered", the motor turns the whole cell body at speeds around 10 Hz. The rotating heart of the motor is a set of rings in the cytoplasmic membrane, about 45 nm in diameter and containing a total of a few hundred molecules of several different proteins. This rotor is surrounded by a ring of 8 to 16 independent torque generators which are anchored to the cell wall and couple the flux of ions to rotation of the rotor.

Swimming is not enough to allow bacteria to navigate, they also need to sense which way to go and a way to arrange that they go there. The basic strategy appears to be common to many different species. Flagellar motors are able to switch spontaneously, for example either reversing direction or switching between "run" and "stop" states. Switches lead to a change in swimming direction and occur stochastically at intervals of several seconds. Navigation is accomplished by modulation of the intervals between changes of

A.T. Skjeltorp and A.V. Belushkin (eds.), Forces, Growth and Form in Soft Condensed Matter: At the Interface between Physics and Biology, 145-164.

direction according to temporal changes in attractant or repellant levels. For example, *Escherichia coli* is propelled by a co-ordinated bundle of several counterclockwise-rotating flagella, driven by motors located at random on the cell surface (Macnab, 1996). When a motor switches to rotation in the opposite direction it is expelled from the bundle and the new pattern of forces on the filament causes it transiently to assume a kinked shape, changing the swimming direction in the process. These events, called tumbles, have been studied for decades, although the details of how motor switches lead to changes in swimming direction have only recently been revealed (Turner et al., 2000). The cell compares conditions sensed within the last second to those over the preceding several seconds, and suppresses tumbles when things are improving (Segall et al., 1986). This strategy is tailored to the physical constraints imposed by the small size of the cell, which loses track of its direction over several seconds due to Brownian Motion and is too small to tell at which of its ends the concentration of a chemical attractant is higher (Berg and Purcell, 1977). In other species the details are slightly different. *Rhodobacter sphaeroides* has a single flagellum which coils up when the motor stops, re-orienting the cell in the process, while other species simply swim in the opposite direction upon motor reversal. The basic strategy of increasing the likelihood of a change of direction when conditions deteriorate and/or reducing it when conditions improve, however, appears to be the same.

Although experiments on tethered bacteria in the 1970s were the first observations of single molecular motors, considerably more is now known about the detailed mechanisms of the ATP-driven motors actin/myosin, kinesin/microtubules and F_1-ATPase. Because the flagellar motor is a large complex, present in the cell in low numbers, and assembled spanning the entire cell envelope, it has not been possible to re-constitute the motor *in vitro*. Because of the size of the motor and its location in the membrane, X-crystallography has to date yielded the structure of only the C-terminal domain of one rotor protein, FliG. In addition, with the pmf rather than ATP hydrolysis as the driving force for rotation, experiments to measure in solution rate constants of chemical steps in the motor cycle are not possible. Similar difficulties exist for nature's other ion-driven motor, the membrane-spanning F_0 motor of ATP-synthase, where rotation has been inferred but never observed directly. Despite these constraints, much has been learned about this fascinating biological electrical rotary motor using a range of electron microscopy, genetic and biophysical techniques. This chapter summarizes what we know.

2. Structure

2.1. COMPONENTS OF THE FLAGELLAR MOTOR

The bacterial flagellum is one of the most complex structures found in bacteria, with 40-50 genes involved in its expression, assembly and control (Macnab, 1996). Not counting the hook and filament, which are part of the propeller rather than the motor itself, nor the LP-ring and the rod, which are the drive-shaft, the flagellar motor contains on the order of 250 protein molecules of 6 different types (Thomas et al., 1999). It is about 50 nanometres in diameter and spans the entire cell envelope from the cytoplasm, through the cytoplasmic membrane, periplasm, cell wall and outer membrane. The molecular weight of the C-ring alone is about 5 Mega Daltons (Francis et al., 1999). The motor includes cytoplasmic membrane proteins and proteins that anchor it to the peptidoglycan cell wall, as well as structural proteins that assemble into rings, a drive shaft and a bushing that carries the shaft through the outer membrane.

Figures 1a and 1b show side-on and end-on views of the rotor obtained by transmission cryo-electron microscopy (cryo-EM) (Francis et al., 1994, Thomas et al., 1999), and figure 1c shows a scanning electron microscope image of rotors visualized by fast-freeze and thin-film metal replica techniques (Khan et al., 1998). Figure 1d shows the stator, a ring of particles left in the membrane after freeze-fracture (Khan et al. 1988). The structure of the entire motor inferred from these images is shown in figure 2 with the locations indicated of various flagellar structural genes that have been identified by immuno-gold labelling and mutational studies. The rotor, drive shaft (rod) and propeller (hook, filament) are shown in white, the stator is shaded.

148

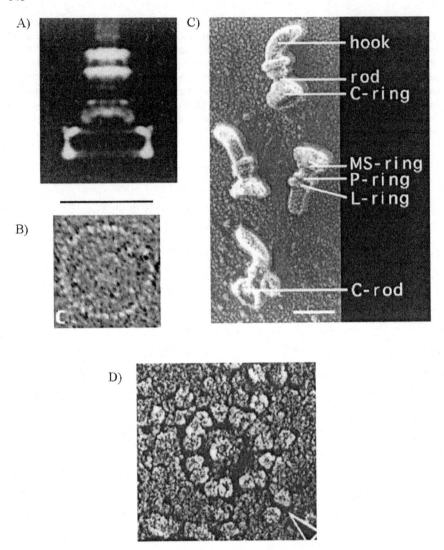

Figure 1: Electron-micrographs of the flagellar motor. a) Side-on projection of a 3-D model of the rotor obtained by alignment and averaging of many single-rotor images. b) End-on view of a single-rotor image of the type used in A. c) Scanning EM of thin-film metal replicas of several rotors. d) Freeze-fracture EM of a ring of stator particles in the cytoplasmic membrane. The dark scale bar is 45 nm and applies to Figures 2a, b and d. The white scale bar in figure 2c is 50 nm.

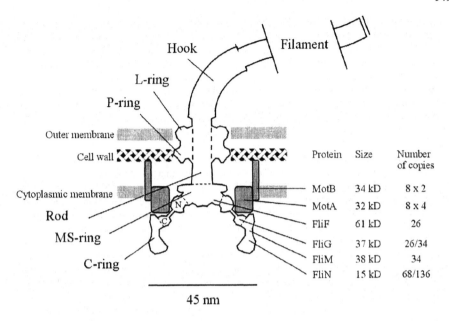

Protein	Size	Number of copies
MotB	34 kD	8 x 2
MotA	32 kD	8 x 4
FliF	61 kD	26
FliG	37 kD	26/34
FliM	38 kD	34
FliN	15 kD	68/136

Figure 2: A schematic diagram of the flagellar motor, showing the various components and the proteins associated with the torque-generating apparatus. Ions drive rotation by flowing across the cytoplasmic membrane at the site of torque generation, the interface between MotA/B and FliG. The structure and mechanism of Na^+ motors are thought to be very similar to those of H^+ motors.

The filament is a helical propeller, typically 5-10μm long, 20 nm wide, and composed of over 10,000 copies of a single protein, flagellin (a detailed review of the structure of the filament and other flagellar components is available, Namba and Vonderviszt, 1997). Torque is believed to be generated by interactions between the stator particles in the cytoplasmic membrane and the protein FliG, located at the junction of the MS- and C-rings. The exact orientation and number of copies of the different proteins in the rotor is not known. Figure 1b shows an end on view of the C-ring in a single rotor. Three such images were seen in the original experiments, two of which had 34-fold, the third 33-fold symmetry (Thomas et al., 1999). Quantitative immunoblot analysis combined with known molecular weights and rotor symmetries indicate a "best-guess" model in which the MS-ring is a hub formed by 26 copies of FliF, 26 or 34 copies of FliG connect to this hub at the N-terminal and generate torque by interaction with the stator at the C-terminal. The rest of the C-ring is attached to FliG, consists of and 34 and 68 or 136 copies of FliM and FliN respectively, and controls motor switching as well as possibly interacting with FliG in torque generation (Thomas et al., 1999, Matthews et al., 1998).

Figure 1d shows a scanning EM image of a ring of studs left in the cytoplasmic membrane after freeze fracture (Khan et al., 1988). The number of studs per ring is variable, but averages around 11. This variability may be the result of freeze fracture pulling some studs into the other face of the membrane, or it may reflect real variability in the number of Mot complexes around the flagellum. The central hole in the rings has the same diameter as the MS-ring and the particle in the middle of the hole matches the diameter of the flagellar rod. Mutants lacking either of the integral membrane proteins MotA or MotB produce apparently normal flagella which do not rotate, and in which no rings of studs can be found. Thus it appears that the ring of studs is the stator of the flagellar motor, and contains at least the proteins MotA and MotB. The model of figure 3 is deduced from the relative sizes of rings in the rotor and in the membrane, and should be treated with caution as there is no direct evidence of the position of the C-ring relative to stator.

Although the Mot proteins show no homology to any previously identified proton translocating proteins, various lines of evidence indicate that the MotA/B stator particles are the proton channels of the motor. Cells that over-express wild-type Mot proteins show reduced growth rates that are attributable to leakage of protons across the cytoplasmic membrane. Paralysed *mot* mutants do not show the reduction in growth rate, and give considerably reduced proton fluxes in artificially energized membrane vesicles made from over-expressing cells (Blair and Berg, 1990, Sharp et al., 1995, Zhou et al., 1998b). Inferred rates of proton flux are low when compared to expected rates based on levels of expression and the energetics of flagellar rotation. This is not surprising as the over-expressed Mot proteins are not incorporated into motors, whereas normal levels of proton flux are expected to depend upon coupling to rotation in complete motors.

The best studied Na^+-driven flagellar motor is the polar motor of the marine bacterium *Vibrio alginolyticus* (Yorimitsu and Homma, 2001). This species swims with a single polar motor in media of low viscosity such as sea water. The proteins PomA and PomB in the Na^+-driven motor appear to be equivalent to MotA and MotB. Genetic fusion and gel purification and filtration studies indicate that Pom proteins are associated into complexes containing 4 molecules of PomA and 2 of PomB, and that each these complexes may represent two functional torque generating units. Early indications are that MotA and MotB in H^+ motors are arranged in the same way (Braun and Blair, 2001), and the results of experiments in which the number of units in H^+ motors is altered while measuring rotation speeds are consistent with complexes containing two functional torque generators.

2.2 ROTOR-STATOR INTERACTIONS

The interactions between rotor and stator that generate torque have not measured directly, but mutational analysis has shed some light on which protein domains are involved. A screen of all conserved acidic and basic residues in the proteins that are involved in torque generation (MotA, MotB, FliG, FliM, FliN) revealed only one that is essential for rotation – aspartate 32 in the membrane-spanning region of MotB (Zhou et al., 1998b). This residue is thought to be part of the proton conducting pathway of the motor. However, 3 charged residues in FliG and two in the cytoplasmic domain of MotA were found that affected rotation when replaced by other residues. Mutations in these residues are highly synergistic, and in particular changes in one protein are able to restore defects in rotation caused by changes in the other protein. The charge of the residues involved is crucial. For example reversing either the charge at position 281 in FliG or at position 98 in MotA impairs rotation considerably more than reversing both charges together. This indicates that the torque generating mechanism relies on electrostatic interactions between the rotor and stator at these residues (Zhou et al., 1998a). Some support for this hypothesis is provided by the crystal structure of the C-terminal domain of FliG (Lloyd et al., 1999), which is consistent with a model in which the key charged residues are on a face that interacts with the stator.

3. Function

A motor is a machine that generates mechanical work from some other form of energy. The natural unit of energy for molecular motors is the thermal energy, kT, where k is Boltzmann's constant and T is the absolute temperature. At room temperature, kT is about 4×10^{-21} J, thus by the principle of Equipartition of Energy each degree of freedom of the motor is in Brownian motion with half this amount of thermal energy. The electrochemical energy released by a proton crossing the bacterial cytoplasmic membrane is equal to the proton charge times the pmf. With a typical pmf of 150 mV, this is about 6 kT. Thus it is clear that molecular motors are subject to Brownian motion in as much as the energy changes involved in their mechanisms are on the same order of magnitude as the thermal energy. There are two complementary approaches to understanding the mechanism of a motor. One is to look at what it's made of and how the parts join up, as in the previous section on motor structure. The other approach is to watch how it moves. New techniques have allowed the observation of single mechanical steps in the cycle that couples the hydrolysis of a single ATP "fuel" molecule to motion in single kinesin, myosin and F_1-ATPase motors (Noji et al, 1997). Similar experiments on the flagellar motor have not yet been successful, due mainly to the small and variable size of the fuel quantum, the fast motor cycle (10-100x faster than known ATP motors) and the difficulty of re-constituting the complex structure in an in-vitro assay that allows the level of control necessary to measure single steps. What has been measured is how torque and speed depend upon pmf, proton flux and each other. These measurements allow models of the mechanism to be tested and are beginning to narrow down the wide range of possible mechanisms that have been proposed.

3.1 A ELECTRIC MOTOR

Aside from its rotary motion, the most striking difference between the flagellar motor and linear molecular motors of eukaryotes is that the flagellar motor is powered by transmembrane ion fluxes rather than ATP hydrolysis. The ion-motive force consists of two components, and may be written as $\Delta p = \Psi + \dfrac{kT}{q}\ln\left(\dfrac{C_i}{C_o}\right)$, where Ψ is the transmembrane voltage (inside minus outside), C_i and C_o are the concentrations of the ions inside and outside the cell respectively, kT is the thermal energy and q the charge of the ion. The electrical component is simply the voltage, Ψ ; the chemical component is due to different ion concentrations on either side of the membrane and is represented by the second term on the right hand side of the equation. Conclusive evidence for a pmf driven motor in bacteria came when it was shown that flagella in cell envelopes containing no cytoplasm ("ghosts"), isolated from E. coli, could rotate if an artificial diffusion potential was generated across the cytoplasmic membrane by the addition of K^+ and valinomycin (Ravid and Eisenbach, 1984). Aside from the identity of the driving ion, Na^+ motors seem to work with essentially the same mechanism as H^+ motors. Lithium supports motor rotation in Na^+ motors (Liu et al, 1990), and deuterium in H^+ motors (Chen and Berg,

2000b), in both cases with normal torque at low speeds but considerably reduced torque at higher speeds. The most compelling evidence that the motors have a common mechanism is the existence of functional chimeras containing components of both types of motor (Yorimitsu and Homma, 2001).

So far there has only been one successful measurement of the proton flux through the flagellar motor. The change in the rate of proton uptake by a population of cells, monitored via changes in the pH of a weakly buffered medium in which the cells were swimming, was measured when flagellar rotation was stopped by adding anti-filament antibody to crosslink the filaments of different motors (Meister et al., 1987). It was deduced that the rotation-dependent proton flux is proportional to rotation rate, and corresponds to the transit of approximately 1200 protons per revolution of the motor. The pmf may be controlled either by using micropipettes (Fung et al 1995), diffusion potentials (Khan et al., 1985) or agents that cause the membrane to leak protons (Gabel and Berg, unpublished). In all cases measured to date, torque is proportional to pmf.

3.2 TORQUE VS SPEED

To understand mechanism of the flagellar motor we need to understand the mechanochemical cycle – what the different states of the motor are and how they couple ion flux and rotation. In the absence of direct measurements on single steps in the cycle, the relationship between torque and speed is one of the best probes of the flagellar mechanism. Data can be compared to the predictions of models with different coupling mechanisms and used to eliminate models that do not fit. The average motor torque and speed in a cell population can be measured via broad peaks in the frequency spectrum of fluctuations in light intensity taken from images of a population of swimming cells (Lowe et al., 1987). By measuring rotation rates in media of different viscosity, an approximately linear decrease in torque between 50 Hz and 100 Hz was found, extrapolating to zero torque at around 110 Hz. Parallel measurements of bundle rotation rates using this technique and linear swimming speeds indicated that the two were directly proportional, allowing swimming speed to be used as an indicator of rotation rate.

However, experiments which control the speed of the motor by altering the viscosity of the medium can only access a limited range of speeds. Using the technique of electrorotation, an external torque can be applied to the motor via the body of a tethered cell, allowing a far wider range of speeds to be examined in a single motor. In particular, the motor can be studied under conditions where it is made to rotate backwards, or forwards at speeds higher than the zero-torque speed, by the electrorotation torque. These regimes are not accessible in cells rotating under their own power. The phenomenon of electrorotation has been known for a long time, but has only recently been applied to studies of the bacterial flagellar motor (Washizu et al., 1993, Berg and Turner, 1993, Berry and Berg, 1998). Electrorotation of bacteria is illustrated in figure 3. Essentially, a rapidly rotating electric field polarizes the cell body and the surrounding medium. Due to the high frequency of rotation (several MHz), the polarization lags behind the electric field, and thus a torque is exerted upon the cell. This torque is proportional to the square of the electric field amplitude.

Even at the relatively high speeds accessible with electrorotation, Reynolds number for a tethered cell is on the order of 10^{-3}, which means that inertial forces are negligible. Therefore the rotation rate of the flagellar motor is always directly proportional to the total torque, with the constant of proportionality being the rotational viscous drag coefficient of whatever is rotating. The torque generated by the flagellar motor was measured by measuring rotation speeds with the same applied torque before and after breaking the motor. Motors were either broken mechanically or de-energized by ionophores or ultraviolet irradiation that killed the cells. After breaking motors did not rotate on their own, and rotated at speeds proportional to the applied torque when electrorotated. The motor torque is thus proportional to the difference in speeds recorded at the same applied torque before and after breaking. Absolute values of torque can be calculated by estimating the viscous drag coefficient of the tethered cell (Meister and Berg, 1987), or alternatively by reference to stall torque measured with a calibrated optical trap (figure 4, Berry and Berg, 1997).

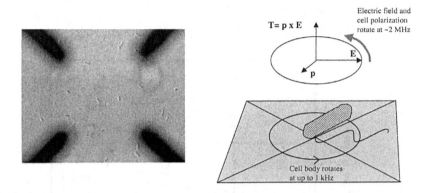

Figure 3: Electrorotation of tethered cells. Tethered cells are attached to a glass coverslip by a single flagellum, so that the motor causes the entire cell body to rotate (right). The gap between the micro-electrodes (left) is 40 μm, and with up to 75 V applied to each opposing pair of electrodes enough torque can be applied to spin tethered cells at speeds up to 1000 Hz in either direction.

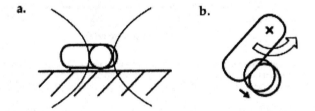

Figure 4: Measuring torque with an optical trap. An optically trapped polystyrene sphere of diameter 1 μm is used to stall a tethered cell. a) side-view, b) top view. The displacement of the bead is a measure of the force exerted upon it, and may be used to calculate the stall torque of the motor.

Figure 5 summarizes torque-speed curves obtained at different temperatures. The most unusual feature, compared to torque-speed curves for other molecular motors, is the wide range of speeds over which the torque is nearly constant. The torque plateau has been confirmed in experiments where high speeds are achieved without electrorotation by attaching small polystyrene beads to truncated flagella (by rapid exchange of media of different viscosity, the speed of rotation of single flagellar motors was varied up to ~300 Hz, Chen and Berg, 2000a). Another notable feature of the torque-speed relationship is the continuity of torque either side of zero speed. This has been confirmed independently by using optical tweezers to stall tethered cells and then to measure the torque they exert either when pushed slowly backwards or allowed to rotate slowly forwards (Berry and Berg, 1997). The result is in contrast to similar measurements on kinesin, in which it was found that single motors could be stalled by a force of 5 pN, but would not slip backwards even when 13 pN was applied (Coppin et al., 1997). The continuity of torque either side of stall in the flagellar motor is evidence against a ratchet-like mechanism in which there is an irreversible step in the motor cycle.

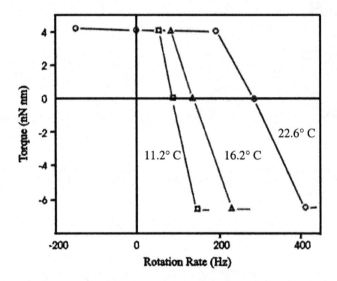

Figure 5: Torque vs speed in the bacterial flagellar motor. The motor generates nearly constant torque up to a certain speed that depends upon temperature. At higher speeds less torque is generated, and at very high speeds the motor resists rotation.

The torque plateau indicates that internal processes in the torque-generating cycle of the motor, (for example the motion of protons through the motor or possibly torque-generating conformational changes) are not rate limiting at these speeds. Rather, the rotation rate is limited mechanically by the load on the motor. At higher speeds torque is much reduced, indicating that internal processes are now rate limiting. At the zero-torque speed these internal processes dissipate all the energy available to the motor, and the output power goes to zero. As an analogy, imagine a person riding a bicycle with one gear and no freewheeling. The "internal processes" are the motions of the rider's legs on the pedals, the output torque is what the rider applies to the back wheel via the chain. At low speeds and high loads, for example riding uphill, the rider can push the pedals with all his force and will produce constant torque. But if he is going very fast downhill, there comes a point where he can only just move his legs around fast enough to keep up with the pedals, and is not pushing at all. Here, all his efforts are being dissipated in the motion of his legs. This interpretation is confirmed by the temperature dependence of the torque-speed curves and also by experiments in which substitutions of non-physiological ions reduced motor torque in swimming cells (fast rotation) but not in tethered cells (slow rotation) (Liu et al., 1990, Deuterium reference as above). The rate-limiting steps at high speed are thermally activated and dependent upon the driving ion, as expected for internal electrochemical processes; but at low speeds the rate limiting steps are independent of temperature and ion type, as expected for mechanical relaxation of a strained torque-generating state. Mechanical equilibrium at low speeds is also consistent with the proportionality of torque vs pmf and flux vs speed. 1200 protons per revolution and a torque plateau of 4000 pN nm at 150 mV correspond to an efficiency close to unity, indicating that the motor operates close to equilibrium in the torque plateau.

3.3 INDEPENDENT TORQUE GENERATORS

Good evidence exists that the particles in stator ring of figure 1d are independent torque generators containing the proteins MotA and MotB. "Resurrection" experiments have been carried out in *E. coli* in which wild-type MotA or MotB is expressed from an inducible plasmid in tethered cells of a non-motile mutant in the same *mot* gene (Block and Berg, 1984, Blair and Berg 1988). These cells initially do not rotate, but start to rotate and then speed up in equal speed increments over the

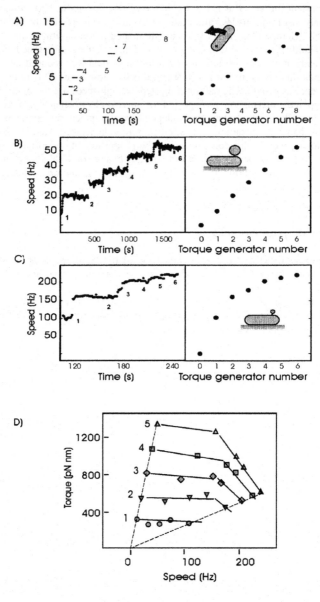

Figure 6: Resurrection of mot mutants in E. coli. Incorporation of torque generating complexes after induction of wild-type mot genes in (a) tethered cells, (b) cells labelled with 1.0 micron beads and (c) cells labelled with 0.3 micron beads. d) Torque-speed relationships for different numbers of complexes from 1 to 5.

course of several minutes as the wild-type protein is expressed and incorporated into the motor (figure 6a). The maximum number of torque generating units per motor remains uncertain, but the most likely explanation of the data is that there are up to sixteen units which come and go in pairs (Fung and Berg, 1995, Berry et al., 1995). This is consistent with the structural model in which a Mot complex contains two torque generating units.

Resurrection experiments with polystyrene beads attached to truncated flagella confirm that torque generating units act independently at high as well as at low speeds (Ryu et al., 2000). Figures 6b and 6c show resurrection of motors attached to beads of diameter 1.0 μm and 0.3 μm respectively. The 1.0 μm data are essentially the same as those from tethered cells. At each step a new Mot complex adds a torque corresponding to 10 Hz, regardless of the other complexes and regardless of the increase in motor speed from 10 Hz for the first step to 60 Hz for the last step. The data with 0.3 μm beads are different. The first step adds 100 Hz, whereas the last step adds far less. After the last step the motor is rotating at 230 Hz, and the observed reduction in the torque generated per unit can be explained by considering the torque speed curve of figure 5, which indicates that the motor at room temperature generates considerably less torque at 230 Hz than it does at 100 Hz. By estimating the drag coefficients of beads of different bead sizes, torque-speed curves can be constructed for different numbers of Mot complexes from 1 to 5 (figure 6d). Comparison with a simple model indicates; that the torque-speed relationship is property of individual torque-generating complexes, that there is no "freewheeling" (complexes have a high duty ratio), that each complex acts independently, and that its rate-limiting transition at low load cannot be speeded up by torque imparted by other units via the rotor. The only interaction between the torque generators appears to be that they share the viscous load and thus increase the speed of the motor compared to a single generator driving the same size polystyrene bead.

3.4 REVERSIBILITY

Another remarkable feature of the flagellar motors of many species is their reversibility. Switches are fast, occurring within 1 ms in single filaments (Kudo et al., 1990), and the motor generates similar torque in either directional mode (Berry et al., 1995). Switches occur at random time intervals, and individual switches of different motors on the same cell are not at all correlated, indicating that the stochastic process controlling motor switching occurs at the level of the individual motor (Ishihara et al., 1983, Macnab and Han, 1983). Switching is controlled by the binding of the active form of the chemotaxis protein CheY to FliM in the C-ring. In the absence of CheY, at room temperature, motors in *E. coli* rotate exclusively counterclockwise (CCW). Active CheY increases the probability of clockwise (CW) over CCW rotation in *E. coli*, presumably by reducing the free energy of the CW state relative to the CCW state. However, CW rotation can be observed in mutants lacking CheY if the temperature is reduced close to 0° C, indicating an entropic contribution to the free energy difference between different rotational states (Turner et al., 1996).

Current indications are that the basic torque generating cycle of the motor is reversible with respect to the direction of proton flux. Berg et al. (1982) found a mutant *streptococcus* which did not switch in response to changes in cytoplasmic pH, but rotated in the opposite direction upon reversal of an artificially generated pmf. Using a voltage clamp technique, Fung and Berg (1995) found that reversing the direction of the pmf in *E. coli* usually caused motors to stop rotating, de-activating upon removal of the normal pmf. In 5 cases out of 17, however, motors reversed direction and rotated a few times before stopping. (It is not clear why *streptococcus* motors did not also de-activate upon reversal of the pmf.). Thus the bacterial flagellar motor appears to be reversible at two levels. First, it possesses two modes which couple proton influx in normally energized cells to rotation in either direction. Second, each mode can probably be driven in either direction depending on the direction of the pmf. However, the effects of reversing the pmf upon the basic mechanism of the flagellar motor are confused, in *E. coli* at least, by the disassembly of motors in the absence of a normal pmf.

3.5 STEPS?

A great deal has been learned about the mechanism of ATP-driven molecular motors by observing single mechanical steps corresponding to the hydrolysis of a single ATP molecules. Kinesin takes steps of 8 nm per ATP hydrolysed, equal to the $\alpha\beta$-tubulin repeat length along the microtubule. F_1-ATPase takes $120°$ steps, corresponding to the 3-fold symmetry of the motor. Based on the 34-fold symmetry of the C-ring, one might expect to see steps of around $11°$ in the rotation of a flagellar motor powered by a single torque-generator. However, if 1200 protons cross the membrane per revolution in a wild-type motor with approximately 10 torque generators, one would expect steps of $3°$ for each proton in a single-generator motor.

Technical difficulties have so far defied all attempts to measure directly steps in the rotation of the flagellar motor, and the possibility remains that the motor mechanism is such that there are no physical steps to be seen. It is, however, possible to infer the stochastic behaviour of the motor from fluctuations in rotation speed. Variance analysis of the time intervals taken by tethered cells to cover a fixed number of revolutions reveals fluctuations that are intrinsic to the flagellar motor mechanism (Berg et al., 1982). Under the assumption that rate-limiting events in the motor cycle occur at random times (a Poisson process), the data indicate that each revolution includes around 400 such events. Contributions to the variance from processes other than motor stepping would make this number an underestimate, while any regularity in the intervals between events would mean that 400 is an overestimate. A more sophisticated experiment, using electrorotation to control cell rotation rates (Samuel and Berg, 1995), excluded the possibility that the variance was due to free rotational Brownian motion superimposed upon a smooth motor rotation. As in the earlier work, these results were consistent with about 400 randomly occurring events per revolution of the motor. When the experiment was repeated in a resurrection strain in which motors contained between 1 and 4 torque generators, the predicted number of events was proportional to the number of generators, with about 50 per generator per revolution (Samuel and Berg, 1996). Whether these "events" correspond to the transit of single protons or to some other chemical or mechanical step in the working cycle of the motor remains to be seen.

4. Models

The level of detail in our current understanding of the structure and function of the bacterial flagellar motor, based on experimental results, leaves quite a lot of room for alternative models of the motor mechanism. Several discussions of competing models may be found

(Berg and Turner, 1993, Berry, 2000), here we give an overview of the different modelling approaches that have been taken and what they can tell us.

Figure 7 shows three different ways in which the flagellar motor could generate torque. In the "turbine" mechanism, protons interact simultaneously with rows of components on both the rotor and the stator. These rows are tilted with respect to one another, and protons flow into the cell by travelling at their point of intersection. Proton influx makes the rotor and stator slide relative to each other so that the intersection of the tilted rows follows the proton as it passes through the

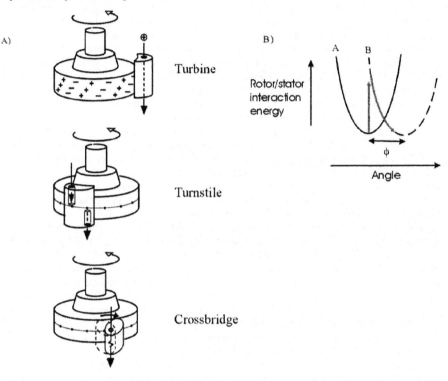

Figure 7: Models of the Flagellar motor mechanism. a) Three different types of model for the mechanism of the flagellar motor. b) Models can be understood in terms of motor states with different rotor-stator interaction energy profiles. For example, a transition from state A to state B as shown by the grey arrow requires the input of chemical energy, which will *be* released as work as the motor in state B relaxes towards its new angle of minimum energy.

motor. The tilted rows on the rotor could be lines of charged residues, so for example alternate negatively and positively charged lines respectively would attract and repel a transient ion, coupling its transit to rotation (Berry, 1993). Alternatively, the lines on the rotor and stator could be residues that each constitute half a binding site for the ion. Binding sites are formed only at the intersection of rotor and stator lines, and thus a transient would carry the intersection with it across the membrane, again leading to rotation (Lauger, 1988). The structure of the C-terminal domain of FliG shows alternating positive and negative charges on the face that interacts with MotA, which could form alternating lines of charges.

In the "turnstile" mechanism, ions are channelled onto the rotor from outside the cell (Meister et al., 1989). The rotor then moves, either under the effect of some force on the charged ions or simply due to thermal fluctuations. Each ion is allowed to pass into the cytoplasm, completing the motor cycle, only after the rotor has moved a certain distance in the right direction. With the ion gone, the rotor is locked to the stator again, and the motor has made one step. This mechanism is favoured for the proton-translocating F_O rotary motor of ATP-synthase, with a conserved aspartate residue in the c-subunit identified as the proton-binding site on the rotor. However, it is worth noting here that extensive mutagenesis of C-ring proteins has failed to identify any possible binding sites on the rotor of the flagellar motor (Zhou et al., 1998a).

The turbine and turnstile models explicitly require the well-defined structure of the flagellar motor and its placement in the cell envelope in the torque-generating process. The "crossbridge" mechanism, on the other hand, is based on the linear motors of eukaryotes where single motor molecules appear to be able to move along their filament substrates, essentially in free solution. The defining element of the crossbridge model is that the stator undergoes a conformational change while bound to the rotor, driven by ion translocation, that generates elastic strain in the crossbridge and causes the rotor to rotate. The rotor must then release the stator so that the conformational change can be reversed without reversing the rotation. Conformational changes in the cytoplasmic domain MotA (containing the key residues that interact with the rotor) have been linked to protonation of the aspartate residue at position 32 in MotB (asp32) (Kojima and Blair, 2001). Conformational changes are inferred from changes in the pattern of proteolysis fragments of MotA caused by mutations in asp32 of MotB, and the link to protonation is based on the observation that the most effective mutations were those that most closely mimic a neutral, protonated aspartate residue at position 32. In addition, changes were blocked by dominant non-rotating mutations of proline 173 in MotA, which also block proton flux though the Mot complex. This has led to a crossbridge model in which protonation neutralizes asp32, leading to conformational changes of MotA that exert torque on the rotor. The same conformational changes should also couple the rotor angle to proton access to asp 32, although there are no structural details to indicate how this might be achieved.

In any of the above schemes, the most likely mechanism for motor switching is a co-ordinated transformation of the C-ring which reverses the handedness of the motor's helical symmetry. In practice, the boundaries between these categories become blurred when a detailed treatment is attempted. Nonetheless, they form useful guide to describing and understanding specific mechanisms, and help to illustrate the similarities and key differences between competing models.

4.1 KINETIC MODELS

Regardless of the conceptual, structural or physical details of a model, an explicit mathematical description is needed in order to yield predictions that can be compared with experimental data. The classical approach, which can be traced to early models of muscle (Huxley and Simmons, 1971), is to define distinct chemical states for the motor, each with a characteristic functional dependence of free energy upon the rotor angle. Thermally activated transitions between these states are allowed at appropriate angles and are driven by proton translocation. Transition rates depend indirectly on external load, which alters the probability of different rotor angles at which, in turn, the transition rates are different. An illustrative example is shown in figure 7b. Energy profiles for two states are shown, with energy minima separated by an angle ϕ. Minima could be the rotor angles at which opposite charges on the rotor and stator align in a turbine model, or alternatively the angle of minimum elastic strain in a crossbridge model. A transition from the energy minimum of state A to state B requires electrochemical energy, and leads to torque generation as the motor in state B relaxes to its new angle of minimum energy. Any model is fully specified by the energy profiles of the motor states, the transition rates between states as functions of rotor angle, and the rotational diffusion coefficient of the rotor.

Once the model is specified, Monte-Carlo simulations of the equation of motion of the rotor coupled with the allowed kinetic transitions can be used to obtain predicted trajectories of the motor, while state occupancies, averaged torques and speeds can be found by numerical solution of the associated reaction-diffusion equations (Elston and Oster, 1997). For a simpler treatment, some authors assume tight coupling between proton flux and rotation and combine chemical transitions and the subsequent mechanical relaxation to the new energy minimum into a single step, with a rate constant that depends explicitly on the mechanical load (Meister et al., 1989). Fitting such a model to experimental torque-speed relationship indicates that the flagellar motor operates with a "powerstroke" mechanism. That is, there is a step in the cycle that converts most of the electrochemical energy of the proton directly into rotation (Berry and Berg, 1998). This is in contrast to "thermal ratchet" models, in which the role of the free energy supplied by the influx of protons is to "save" thermal fluctuations in a certain direction rather than to create directly a torque-generating state. As more experimental data is collected on the performance of the flagellar motor, models will need to be extended and adapted to interpret these data and to narrow down the range of possible mechanisms.

5. References

1. Armitage, J. P. 1999. Bacterial tactic responses. *Adv Microb Physiol.* 41:229-89.

2. Berg, H.C., and E.M. Purcell. 1977. Physics of chemoreception. *Biophys. J.* 20:193-219.

3. Berg, H.C., Manson, M.D., and Conley, M.P. (1982). Dynamics and energetics of flagellar rotation in bacteria. In Prokaryotic and eukaryotic flagella. W.B. Amos and J.G. Duckett, eds. (Cambridge: Cambridge University Press), pp. 1-31.

4. Berg, H.C. and Turner, L. (1993). Torque generated by the flagellar motor of *Escherichia coli*. *Biophys J.* 65: 2201-2216.

5. Berry, R.M. (1993). Torque and switching in the bacterial flagellar motor: an electrostatic model. *Biophys. J.* .64:961-973.

6. Berry, R.M., Turner, L and Berg, H.C. (1995). Mechanical limits of bacterial flagellar motors probed by electrorotation. *Biophys. J.* 69:280-286.

7. Berry, R.M. and Berg H.C. (1997) Absence of a barrier to backwards rotation in the bacterial flagellar motor, demonstrated with optical tweezers. *Proc. Natl. Acad. Sci. USA.* 94:14433-14437.

8. Berry, R.M. and Berg H.C. (1998). Torque generation by the flagellar motor of *Escherichia coli* while driven backwards. *Biophys J.* 76:580-587.

9. Berry, R.M. and Armitage, J.P. (1999). The bacterial flagella motor. *Adv. Microb. Physiol.* 41:291-337.

10. Berry, R.M. (2000) Theories of Rotary Motors. *Philos. Trans. R. Soc. Lond. B. Biol. Sci.* 355:503-511.

162

11. Buechner, M., Delcour, A.H., Martinac, B., Adler, J., and Kung, C. (1990). Ion channel activities in the *Escherichia coli* outer membrane. *Biochim.Biophys.Acta.* 1024: 111-121.

12. Blair, D.F. and Berg, H.C. (1988). Restoration of torque in defective flagellar motors. *Science.* 242: 1678-1681.

13. Blair, D.F. and Berg, H.C. (1990). The MotA protein of E. coli is a proton-conducting component of the flagellar motor. *Cell* 60: 439-449.

14. Blair, D. F. and Berg, H. C. (1991). Mutations in the MotA protein of *Escherichia coli* reveal domains critical for proton conduction. *J.Mol.Biol.* 221: 1433-1442.

15. Blair D.F. (1995) How bacteria sense and swim. *Annu. Rev. Microbiol.* 49: 489–522.

16. Block, S.M. and Berg, H.C. (1984). Successive incorporation of force generating units in the bacterial rotary motor. *Nature.* 309: 470-472.

17. Braun T. F., and D. F. Blair. 2001. Targeted disulfide cross-linking of the MotB protein of *Escherichia coli.* *Biochemistry* 40:13051-13059.

18. Chen X, Berg HC. 2000a. Torque-speed relationship of the flagellar rotary motor of *Escherichia coli.* *Biophys. J.* 78:1036-1041.

19. Chen X, Berg HC. 2000b. Solvent-isotope and pH effects on flagellar rotation in *Escherichia coli.* *Biophys. J.* 78:2280-2284.

20. Coppin C. M., Pierce D. W., Hsu L., and R. D. Vale. 1997. The load dependence of kinesin's mechanical cycle. *Proc. Natl. Acad. Sci. U S A.* 94:8539-8544.

21. Elston, T.C. and Oster, G. (1997). Protein turbines. I: The bacterial flagellar motor. *Biophys. J.* 73: 703-721.

22. Francis, N. R., Sosinsky, G. E., Thomas, D. and D.J. DeRosier. 1994. Isolation, Characterization and structure of bacterial flagellar motors containing the switch complex. *J. Mol. Biol.* 235:1261-1270.

23. Fung, D.C. and Berg, H.C. (1995). Powering the flagellar motor of *Escherichia coli* with an external voltage source. *Nature.* 375: 809-812

24. Hirota, N., and Y. Imae. 1983. Na+-driven flagellar motors of an alkalophilic *bacillus* strain YN-1. *J. Biol. Chem.* 258:10577-10581.

25. Huxley, A.F. and Simmons, R.M. (1971). Proposed mechanism of force generation in striated muscle. *Nature.* 233:533-538.

26. Ishihara, A., Segall, J.E., Block, S.M., and Berg, H.C. (1983). Coordination of flagella on filamentous cells of *Escherichia coli.* *J.Bacteriol.* 155: 228-237.

27. Jones, C.J., Macnab, R.M., Okino, H., and Aizawa, S.-I. (1990). Stoichiometric analysis of the flagellar hook-(basal-body) complex of *Salmonella typhimurium.* *J.Mol.Biol.* 212: 377-387.

28. Khan, S., Meister, M., and Berg, H.C. (1985). Constraints on flagellar rotation. *J.Mol.Biol.* 184: 645-656.

29. Khan, S., Dapice, M., and Reese, T.S. (1988). Effects of mot gene expression on the structure of the flagellar motor. *J. Mol. Biol.* 202: 575-584.

30. Khan, S., Zhao, R., and T. S. Reese. (1998). Architectural features of the *Salmonella typhimurium* flagellar motor switch revealed by disrupted C-rings. *J. Struct. Biol.* 122:311-319.

31. Kojima, S. and D. F. Blair. 2001. Conformational change in the stator of the bacterial flagellar motor. *Biochemistry.* 40: 13041-13050.

32. Kudo, S., Magariyama, Y., and Aizawa, S.-I. (1990). Abrupt changes in flagellar rotation observed by laser dark-field microscopy. *Nature.* 346: 677-680.

33. Lauger, P. (1988). Torque and rotation rate of the flagellar motor. *Biophys. J.* 53: 53-65.

34. Liu, J.Z., Dapice, M., and Khan, S. (1990). Ion selectivity of the *Vibrio alginolyticus* flagellar motor. *J.Bacteriol.* 172: 5236-5244.

35. Lowe, G., Meister, M., and Berg, H.C. (1987). Rapid rotation of flagellar bundles in swimming bacteria. *Nature.* 325: 637-640.

36. Lloyd, S. A., Whitby, F. G., Blair, D. F., and C. P. Hill. 1999. Structure of the C-terminal domain of FliG, a component of the rotor in the bacterial flagellar motor. *Nature.* 400:472-475.

37. Macnab, R.M. and Han, D.P. (1983). Asynchronous switching of flagellar motors on a single cell. *Cell.* 32: 109-117.

38. Macnab, R.M. (1996). Flagella and motility. In *Escherichia coli* and *Salmonella*:Cellular and Molecular Biology. F.C. Neidhardt, R. Curtiss, I, J.L. Ingraham, E.C.C. Lin, G. Lowe, B. Magasanik, W.S. Reznikoff, M. Riley, M. Schaechter, and H.E. Umbarger, eds. (Washington: ASM), pp. 123-145.

39. Matthews, M. A. A., Tang, H. L., and D. F. Blair. 1998. Domain analysis of the FliM protein of *Escherichia coli*. *J. Bacteriol.* 180:5580-5590.

40. McCarter, L.L. (1994a). MotX, the channel component of the sodium-type flagellar motor. *J.Bacteriol.* 176: 5988-5998.

41. McCarter, L.L. (1994b). MotY, a component of the sodium-type flagellar motor. *J.Bacteriol.* 176: 4219-4225.

42. Meister, M., Lowe, G., and Berg, H.C. (1987). The proton flux through the bacterial flagellar motor. *Cell.* 49: 643-650.

43. Meister, M. and Berg, H.C. (1987). The stall torque of the bacterial flagellar motor. *Biophys. J.* 52: 413-419.

44. Meister, M., Caplan, S.R., and Berg, H.C. (1989). Dynamics of a tightly coupled mechanism for flagellar rotation. *Biophys. J.* 55: 905-914.

45. Namba, K., and F. Vonderviszt. 1997. Molecular architecture of bacterial flagellum. *Q. Rev. Biophys.* 30:1-65.

46. Noji, H., Yasuda, R., Yoshida, M., and Kinosita, K. (1997). Direct observation of the rotation of F1-ATPase. *Nature* 386: 299-302.

47. Ravid, S. and Eisenbach, M. (1984). Minimal requirements for rotation of bacterial flagella. *J. Bacteriol.* 158: 1208-1210.

48. Ryu, W.S., Berry, R.M. and Berg, H.C. (2000) Torque generating units of the flagellar motor of *Escherichia coli* have a high duty ratio. *Nature.* 403:444-447.

49. Samuel, A.D.T. and Berg, H.C. (1995). Fluctuation analysis of rotational speeds of the bacterial flagellar motor. *Proc.Natl.Acad.Sci.USA* 92: 3502-3506.

50. Samuel AD, Berg HC. 1996. Torque-generating units of the bacterial flagellar motor step independently. *Biophys. J.* 71:918-923.

51. Segall, J.E, S.M. Block, and H.C. Berg. 1986. Temporal comparisons in bacterial chemotaxis. *Proc. Natl. Acad. Sci. USA.* 83:8987-8991.

52. Sharp, L.L., Zhou, J., and Blair, D.F. (1995). Tryptophan-scanning mutagenesis of MotB, an integral membrane protein essential for flagellar rotation in *Escherichia coli*. *Biochemistry* 34: 9166-9171.

53. Shioi, J-I., Matsuura, S., and Y. Imae. 1980. Quantitative measurements of proton motive force and motility in *Bacillus subtilis*. *J. Bacteriol.* 144:891-897.

54. Sosinsky, G.E., Francis, N.R., DeRosier, D.J., Wall, J.S., Simon, M.N. and Hainfeld, J (1992). Mass determination and estimation of subunit stoichiometry of the bacterial hook basal-body flagellar complex of *Salmonella typhimurium*. By scanning transmission electron microscopy. *Proc. Natl. Acad. Sci.* 89: 4801-4805.

55. Stock, J.B. and Surette, M.G. (1996). Chemotaxis. In *Escherichia coli* and *Salmonella*:Cellular and Molecular Biology. F.C. Neidhardt, R. Curtiss, I, J.L. Ingraham, E.C.C. Lin, K.B. Low, B. Magasanik, W.S. Reznikoff, M. Riley, M. Schaechter, and H.E. Umbarger, eds. (Washington: ASM), pp. 1103-1129.

56. Thomas, D. R., Morgan, D. G., and D. J. DeRosier. 1999. Rotational symmetry of the C ring and a mechanism for the flagellar rotary motor. *Proc. Natl. Acad. Sci. USA* 96:10134-10139.

57. Turner, L., Caplan, S.R., and Berg, H.C. (1996). Temperature-induced switching of the bacterial flagellar motor. *Biophys. J.* 71: 2227-2233.

58. Turner, L., Ryu W. S, and H. C. Berg. 2000. Real-time imaging of fluorescent flagellar filaments. *J Bacteriol.* 182:2793-801.

59. Washizu, M., Kurahashi, Y., Iochi, H., Kurosawa, O., Aizawa, S.-I., Kudo, S., Magariyama, Y., and Hotani, H. (1993). Dielectrophoretic measurement of bacterial motor characteristics. *IEEE Trans.Ind.Applications.* 29: 286-294.

60. Yorimitsu, T., and M. Homma. 2001. Na⁺-driven flagellar motor of *Vibrio*. *Biochim. Biophys. Acta.* 1505: 82-93.

61. Zhou, J., Lloyd, S.A., and Blair, D.F. (1998a). Electrostatic interactions between rotor and stator in the bacterial flagellar motor. *Proc.Natl.Acad.Sci.U.S.A.* 95: 6436-6441.

62. Zhou J, Sharp LL, Tang HL, Lloyd SA, Billings S, Braun TF, and DF Blair. 1998b. Function of protonatable residues in the flagellar motor of *Escherichia coli*: a critical role for Asp 32 of MotB. *J. Bacteriol.* 180:2729-2735.

SELF-ASSEMBLY AND DYNAMICS OF MAGNETIC HOLES

A.T. SKJELTORP[1,2], J. AKSELVOLL[1,2],
K.de LANGE KRISTIANSEN[1,2], G. HELGESEN[1], R. TOUSSAINT[2,3],
E.G. FLEKKØY[2], and J. CERNAK[4]
[1]*Institute for Energy Technology, POB 40, NO-2027 Kjeller, Norway*
[2]*Department of Physics, University of Oslo, NO-0316 Oslo, Norway*
[3]*Department of Physics, NTNU, NO-7491 Trondheim, Norway*
[4]*University P.J. Safarik Kosice, Biophysics Department, Jesenna 5, SK-043 54 Kosice, Slovak Republic*

Abstract

Nonmagnetic particles in magnetized ferrofluids have been denoted magnetic holes and are in many ways ideal model systems to test various forms of particle self assembly and dynamics. Some case studies to be reviewed here include:
- Chaining of magnetic holes
- Braid theory and Zipf relation used in dynamics of magnetic microparticles
- Interactions of magnetic holes in ferrofluid layers

The objectives of these works have been to find simple characterizations of complex behavior of particles with dipolar interactions.

1. Ferrofluids

Figure 1 shows the characteristic features of ideal ferrofluids on different length scales. Ferrofluids consist of coated magnetic nanoparticles dispersed in a carrier liquid. The nanoparticles are so small that they contain only one magnetic domain, i.e. at this length scale it is not energetically favorable to break up into domains as in ordinary bulk magnets. Ferrofluids are in fact an early success story in the commercialization of nanotechnology [1]. In the 1970s, ferrofluids were adopted by the disk drive industry as near-zero friction bearings. Today, ferrofluid bearings are a key component in greatly reducing the incidence of hard-disk failure, and there is also a wide range of other ferrofluid applications. Ferrofluids are often denoted magneto-rheological fluids (MRF) as they may exhibit fast, strong and reversible changes in their rheological properties when a magnetic field is applied. MRF are similar to electro-rheological fluids, but normally much stronger, and more stable and easier to use. Applications of MRF include clutches, damping systems in passenger vehicles, brakes, controllable friction damper that decreases the noise and vibration in washing machines, and seismic mitigation MRF damping systems protecting buildings and bridges from earthquakes and windstorms [2]. Ferrofluids exhibit also many exotic macroscopic phenomena as exemplified in Figure 2.

A.T. Skjeltorp and A.V. Belushkin (eds.), Forces, Growth and Form in Soft Condensed Matter: At the Interface between Physics and Biology, 165-179.

166

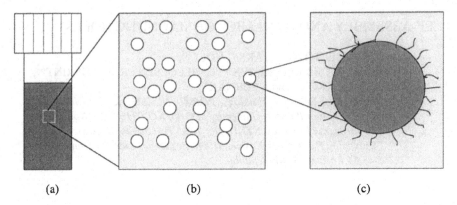

(a)   (b)   (c)

Figure 1. Schematic picture of ferrofluid on three length scales:
a: On a macroscopic length scale, it resembles an ordinary liquid with a uniform magnetization.
b: On a colloidal length scale, solid nanoparticles dispersed in a liquid.
c: Each particle consists of a single domain magnetic core, e.g. iron oxide, and a surface grafted with polymer chains, particle size ~ 10 nm.

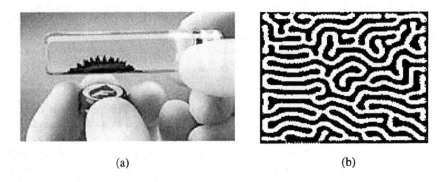

(a)   (b)

Figure 2. (a) Surface instability of ferrofluid subject to an external magnetic field (http://www.ferrotec.com/) and (b) a ferrofluid meander in a thin layer.

2. Magnetic Holes

Monodisperse polystyrene spheres dispersed in ferrofluid provide a convenient model system for the study of various order-disorder phenomena [3]. The basis for this is that the spheres displace ferrofluid and behave as magnetic holes [4] with effective moments equal to the total moment of the displaced ferrofluid. The spheres are much larger (1-100 μm) than the magnetic particles (typically 0.01 μm) in the ferrofluid and the spheres therefore move around in an approximately uniform magnetic background. By confining the spheres and ferrofluid between closely spaced microscope slides an essentially two-dimensional many-body system of interacting particles is obtained. This offers the possibilities of observing directly through a microscope a wide range of nonlinear dynamic phenomena and collective processes, as they are easy to produce and to manipulate with external magnetic fields. A simplifying feature of magnetic holes is that their magnetic moments are collinear with an external field at any field strength. This is in contrast with magnetic particles dispersed in non-magnetic fluids where random orientation of the magnetic moments complicates the theoretical treatment of the dynamic and static properties of the particles.

The basic principle for magnetic holes is shown in Figure 3. It is in some sense a magnetic analogue of Archimedes' principle. When a non-magnetic particle is dispersed in a magnetized ferrofluid ($H > 0$), the void produced by the particle possesses an effective magnetic moment M_V equal in size but opposite in direction to the magnetic moment of the displaced fluid

$$M_V = -V\chi_{eff}H,$$ (1)

where V is the volume of the sphere and $\chi_{eff} = 3\chi/(3+2\chi)$ is the effective volume susceptibility of the ferrofluid. The interaction energy between two spheres with a centre-to-centre separation d is given approximately by the dipolar interaction

$$U = \frac{M_v^2(1 - 3\cos^2\theta)}{d^3}.$$ (2)

Here, θ is the angle between the line connecting the centres of the spheres and the direction of the field. Figure 3 illustrates that if the centers of two holes are collinear, they will attract each other, while two holes side by side will repel. Typical examples of structures formed for the two field orientations are shown in Figure 4.

A detailed description of the interaction between the spheres in a lattice is quite complicated. Since the dipolar interaction is of relatively long range, the direct particle interaction goes far beyond the nearest neighbours. In addition, there is an indirect particle-particle interaction mediated via the glass plates confining the system. This "image dipole" effect is caused by the change in the magnetic permeability across the glass plates. Thus the spheres also interact with their image dipoles situated at the opposite side of the ferrofluid-glass interface. This effect causes the lattice to be situated precisely midway between the upper and lower glass plate. As the plate separation is typically 50-100% larger than the diameter of the spheres, the dipole image contribution is relatively small (typically 10% or less) compared with the interaction energy between the real dipoles. Even in the presence of the dipole images, this experimental system may still be considered to be two-dimensional.

It is possible to obtain a thermodynamic system by using small spheres (diameter less than 2 μm) since Brownian motion introduces fluctuations into the system. The controlling parameter for the stability of the structure formation is the ratio of the dipolar energy to the thermal energy:

$$\Gamma = \frac{M_v^2}{d^3 k_B T}$$ (3)

where d is the (centre-to-centre) separation of the spheres and k_B is Boltzmann's constant.

The ability to design and modify the effective interactions in this system enables studies of various phenomena discussed in the next sections.

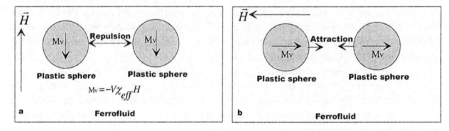

Figure 3. The principle of a magnetic hole in a soft magnet as discussed in the text. (a) Two holes side by side will repel each other. (b) Two holes with the centers collinear with the field lines will attract each other.

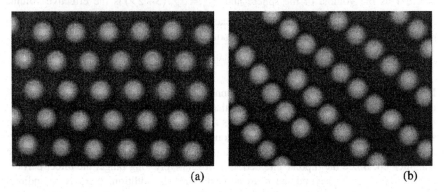

Figure 4. Structures of 10 μm diameter spheres formed by a magnetic field (a) normal to the layer and (b) parallel to the layer.

3. Aggregation and Chaining of Magnetic Holes

Experimental studies of field induced colloidal aggregation have been carried out earlier, e.g., with paramagnetic microspheres [5,6], magnetic nano-particles [7,8], and electric field-induced association of dielectric particles [9]. The results have essentially confirmed the scaling behavior of the mean cluster size as a function of time [10-12] and it has been possible to scale the temporal size distribution of clusters [5,8,13] into a single universal curve as predicted by dynamic scaling theory [14].

In the present work we have studied the chain formation of non-magnetic microspheres [15] dispersed in thin layer of ferrofluid [16] induced by external magnetic fields, see Figure 5.

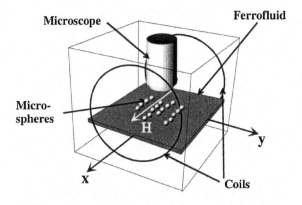

Figure 5. Schematic experimental set-up.

The main advantage of this system is the possibility to tune the strength of the particle-particle interactions. In our case we can create experimental conditions that are close to an ideal point-dipole system, i.e, the particles are spherical, monodisperse, and their resulting induced magnetic moments are oriented in the direction of the external magnetic field. In order to study the importance of dipole-dipole interaction and Brownian motion relative to non-Brownian ballistic drift, we used microspheres with different diameters d = 1.9, 4.0, and 14 µm. In zero external fields the diffusive Brownian motion of the 1.9 µm spheres is clearly visible in the microscope. However, the diffusion of the 14 µm spheres can only be seen by comparing images taken at typically 30 s time lap. The experiments were done with low volume fractions of microspheres, corresponding to coverage of a few percent. Magnetic fields in the range H = 4 – 16 Oe were used. The combination of these fields with the three particle sizes gave rise to values of Γ in the range 8 - 10^4. A typical example of the aggregates that were formed is shown in Figure 6.

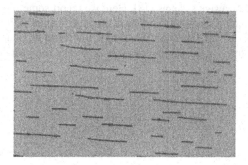

Figure 6. An example of the straight aggregates formed by 4 µm spheres in 10 Oe magnetic field after about 2 hours.

170

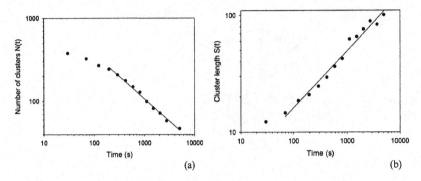

Figure 7. (a) The number of clusters and (b) cluster length versus time for aggregation of 10 μm spheres in a 10 Oe magnetic field. The solid lines are fits to a power law with exponents (a) $z' = 0.53$ and (b) $z = 0.50$.

In order to characterize the aggregation process in more detail, the length s of any cluster was determined at different time intervals t relative to the initial time $t = 0$ when the field was turned on. The time dependences of the number of clusters $N(t)$ and mean cluster size (length) $S(t)$ for a typical experiment with $d = 4$ μm spheres are shown in Figure 7. We see that the data asymptotically follow the power laws $N(t) \sim t^{-z'}$ and $S(t) \sim t^{z}$ with scaling exponents $z' = 0.53$ and $z = 0.50$ for $t > 100$ s. Using the dynamical scaling relations for cluster-cluster aggregation [12,14], it was possible to collapse all the cluster size distributions $n_s(t)$ to a scaling function curve [17].

4. Braid Theory and Zipf Relation used in Dynamics of Magnetic Microparticles

Cooperative behavior of interacting microparticles is of great interest both from a fundamental and a practical point of view. The rank-ordering statistics is one way of analyzing such collective diffusive systems [18]. Here we report some results from our studies of a colloidal system of magnetic holes with an experimental set-up like that in Figure 5. Intricate motions of particles in a plane can be illustrated as entangled lines in three-dimensional space-time, Figure 8, and braid theory gives a compact description of these lines [19,20].

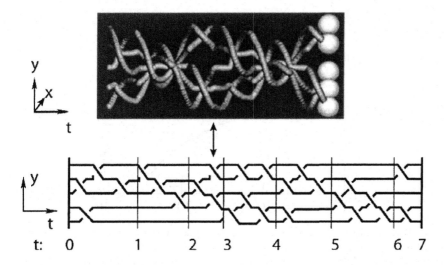

Figure 8. The two-dimensional motion of 5 microspheres is extended to three-dimensional space-time with the trace of each particle in time (upper). Projecting onto (y,t)-axis gives the braid for the motion (lower).

The magnetic holes system consists of 50 μm diameter diamagnetic, polystyrene micro-spheres [15] immersed in a thin layer of ferrofluid [16]. In an external planar, elliptically polarized rotating magnetic field H a microsphere introduces a magnetic hole with apparent magnetic moment given by Eq. (1). The motions of the microspheres can easily be simulated [7] by considering, to first order, the magnetic dipole-dipole interaction between each pair of microspheres and the viscous force on each microsphere.

By varying the angular velocity of the magnetic field and its anisotropy, we observe a rich diversity of patterns. Here we will focus on non-ordered behavior. The structure and complexity of the braids formed by the traces of the moving particles can be captured by calculating a few characteristic numbers. One of these is found by transforming the braid into a positive permutation braid [22] and use that as a measure of the systems dynamical mode. Then we apply rank-ordering statistics on these permutation braids, i.e., count all the different modes and rank them. Figure 9 shows a plot of the frequency of occurrence of the permutation braid $\phi(r)$ vs. rank r. This graph show a power law dependence, according to the Zipf relation [23] $\phi(r) \sim r^{\gamma}$ with $\gamma = 1.2$ for about two decades in r. Attempts to understand the origin of this relation were connected to the structure of hierarchical systems [24].

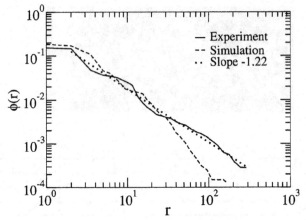

Figure 9. The relative occurrences of braid permutations $\Phi(r)$ against its rank r found in the dynamics of 7 microspheres.

Another characteristic number is the writhe $Wr(t)$. It is equal to the number of over-crossings minus the number of under-crossings, starting with value zero at time zero. Here the time t is measured in units of half periods of the external rotating magnetic field. One measure of the fluctuation in the braid pattern is the variation of the writhe around its mean:

$$\delta Wr(t,\tau) = Wr(\tau + t) - Wr(\tau) - t\overline{\delta Wr}, \quad (4)$$

where $\overline{\delta Wr} = \frac{1}{N}\sum_{t=1}^{N}\left[Wr(t) - Wr(t-1)\right]$. The variance $\sigma^2(t)$ is then:

$$\sigma^2(t) = \left\langle [\delta Wr(t,\tau) - \langle \delta Wr(t,\tau)\rangle]^2 \right\rangle. \quad (5)$$

For diffusive processes: $\sigma^2(t) \sim t^\beta$, with $\beta = 1$ for ordinary diffusion. When β is smaller or larger than 1, the diffusion is anomalous and is denoted as subdiffusion and superdiffusion, respectively.

We have found that the variance of the writhe fluctuations in experiments behave as a power law as shown in Figure 10, with $\beta = 1.66$, clearly indicating anomalous diffusion. Simulations based on the simplest model with magnetic and viscous forces showed good agreement with $\beta = 1.82$. The value of β found in theoretical work [25] using generalized Langevin and Fokker-Planck equations in combination with earlier experimental results [26] agrees well with this experimental result.

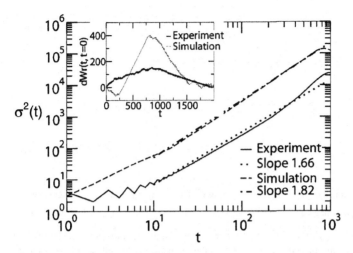

Figure 10. The variances $\sigma^2(t)$ of the writhe fluctuations (Eq. 5) for the dynamics of 7 microspheres. The inset illustrates how the writhe of the system evolves with time.

5. Interactions of Magnetic Holes in Ferrofluid Layers

So far there has been no satisfactory theoretical description of the effective interactions between magnetic holes in rotating magnetic fields composed of a high frequency rotating inplane component and a constant normal one, and the existence of the observed stable configurations of holes with a finite separation distance [27] has remained unexplained.

Focusing on the boundary conditions of the magnetic fields along the confining plates, we have derived analytically the pair interaction potential in such oscillating fields, and demonstrated for a wide range of conditions the existence of a secondary minimum whose position depends continuously on the ratio β between the out of plane $H_\perp$ and inplane $H_\parallel$ field components [28]. We compare this theory with experiments where the motion of a pair of particles (holes) is followed.

In presence of a far-range field $\vec{H}$ in a ferrofluid of susceptibility χ, each hole generates a dipolar perturbation of dipolar moment equal to the opposite of the displaced ferrofluid, $\vec{\sigma} = -V\chi_{\it eff}\vec{H}$, where V is the volume of the hole, and $\chi_{\it eff} = 3\chi/(3+2\chi)$ is an effective susceptibility including a demagnetization factor rendering for the boundary conditions of the magnetic field along the spherical particle boundary [4,29]. The susceptibility contrast between the ferrofluid and the two plane nonmagnetic confining plates leads to a different dipolar field perturbation than the infinite medium expression. According to the image method [30], the boundary conditions for the magnetic field along the plates are fulfilled with an addition of an

174

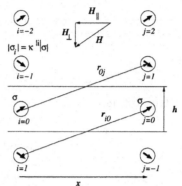

Figure 11. Pair of non magnetic particles in a ferrofluid layer, viewed along the confinement plates, and the series of dipole images accounting for the boundary conditions of the magnetic field along the plates.

infinite series of dipole images to the infinite space expression of the dipolar field emitted in an unbounded medium. The images are constructed as mirror images in the plane boundaries of the initial dipoles or of some previous image, multiplying the magnitude of the dipole at each mirror symmetry operation by an attenuation factor $\kappa = \chi /(\chi + 2)$ - see Figure 11.

The instantaneous interaction potential between a pair of confined holes can then, similarly to Eq. (2), be expressed as

$$U = \frac{\mu}{8\pi} \sum_{i \neq j} \left[\frac{\vec{\sigma}_i \cdot \vec{\sigma}_j}{r_{ij}^3} - 3 \frac{\left(\vec{\sigma}_i \cdot \vec{r}_{ij}\right)\left(\vec{\sigma}_j \cdot \vec{r}_{ij}\right)}{r_{ij}^5} \right], \qquad (6)$$

where μ is the ferrofluid's permability. The i-index runs over both the source and image dipoles, and the j-index runs only over the two source dipoles. A detailed analysis of the above shows that the dominant effect for the force components normal to the plates, is the interaction between a particle and its own mirror images, which stabilizes the particles midway between the plates.

Decomposing the instantaneous field in its inplane and normal components $\overrightarrow{H}_\perp$ and $\overrightarrow{H}_\parallel$, we define the ratio of their magnitudes $\beta = H_\perp / H_\parallel$, the particle diameter and interplate separation respectively as a and h, and the scaled separation as $x = r / h$. At the field rotation frequencies used here, $f = 10 - 100$ Hz, which exceeds the inverse viscous relaxation time, the motion of the holes can be neglected during a field rotation, and an effective, time-averaged interaction potential can be obtained by averaging over a full rotation of the magnetic field, while the slowly varying separation vector is maintained constant. This leads to the dimensionless effective interaction potential

$$u(x) \equiv \frac{144h^3\overline{U}}{\mu\pi a^6 \chi_e^2 H_\parallel^2} = \sum_{l=-\infty}^{+\infty} \kappa^{|l|} \left[\frac{1+(-1)^{|l|}\beta^2}{(x^2+l^2)^{3/2}} - 3\frac{(-1)^{|l|}l^2\beta^2 + y^2/2}{(x^2+l^2)^{5/2}} \right]. \quad (7)$$

The term $l = 0$ corresponds to the source-source interaction term, already used in previous studies [4,27], the others to interactions between a particle and the images of the other one. For all existing ferrofluids, κ is sufficiently smaller than unity so that the prefactor $\kappa^{|l|}$ ensures that the three first images are enough to get a relative precision better than 1% for the potential and its derivatives.

For the micrometer sized particles used here, inertial terms can be neglected, and a characteristic viscous relaxation time can be obtained as the time to separate two particles initially in contact by a distance equal to their size. Balancing a Stokes drag with the magnetic interaction forces derived from the potential above and retaining the main term in Eq.(7), leads to the estimate $T_c = 144\eta / \mu\chi_e^2 H^2 \approx 5s$, for the ferrofluid ($\eta = 9\cdot 10^{-3}$ Pa.s, $\mu = \mu_0(1+\chi)$, $\chi = 1.9$, $\mu_0 = 4\pi\cdot 10^{-7}$ H.m^{-1}) and a typical field $H = 10$ Oe.

For a given ferrofluid and field, this central potential can be of four possible types as illustrated in Figure 12. At low normal field $\beta < \beta_m$, the interactions are purely attractive up to contact; at higher ones $\beta_m < \beta < \beta_c$, a secondary minimum at finite distance appears, later on in a regime $\beta_c < \beta < \beta_u$ this minimum becomes the only one, and ultimately interactions are purely repulsive for $\beta_u < \beta$.

From Eq. (7), the separating values of β can be shown to be $\beta_c = 1/\sqrt{2}$, and $\beta_u = \beta_c(1+\kappa)/(1-\kappa)$ which is a growing function of the susceptibility. β_m is a function of the susceptibility which decreases regularly from β_c at zero susceptibility to 0 at an infinite one. For the ferrofluid used in the present experiments $\beta_m \approx 0.55$ and $\beta_u \approx 2.05$. Neglecting the susceptibility contrast along the plates, i.e., terms with $l \neq 0$ in Eq. (7), would correspond to the limiting case $\kappa = 0$, where these three separating values merge, and the interactions are either purely attractive or repulsive. The presence of this susceptibility contrast is thus essential to trap the particles at a given equilibrium distance in this type of field.

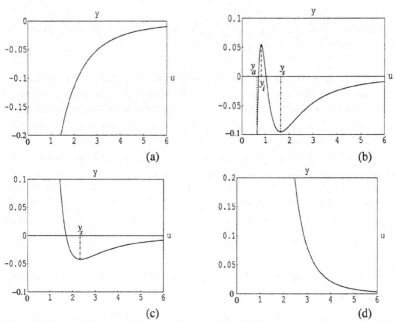

Figure 12. Four possible types of the interaction potential: (a) $\beta<\beta_m$ purely attractive interactions, (b) $\beta_m<\beta<\beta_c$ interactions with secondary minimum, (c) $\beta_c<\beta<\beta_u$ interactions with single equilibrium position at finite distance, (d) $\beta_u<\beta$ purely repulsive interactions.

The motion of pairs of particles initially in contact in a given field was recorded using an optical microscope with an attached video camera. The experiments where carried out using 50 μm diameter nonmagnetic polystyrene particles produced by Ugelstad's technique [15], in a kerozene-based ferrofluid [16]. The confining cell consisted of two glass plates kept at 70 μm separation by using a few 70 μm diameter spheres as spacers. The cell was placed inside three pairs of coils, and the particle motion was recorded and digitized from the microscopy data. The particles were initially brought into contact by applying a fast oscillating, purely inplane field, and at time $t = 0$ the constant normal field component was added. The particle pair observed was separated from any other particle or spacer by more than 20 diameters, in order to avoid unwanted perturbations. A typical record of the scaled distance as function of the scaled time, obtained for a field $H_{\text{I}} = 14.2$ Oe at $\beta = 0.8$ is shown in Figure 13.

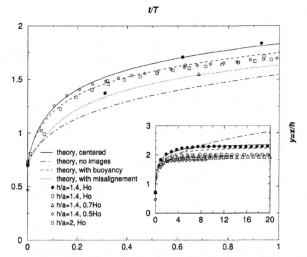

Figure 13. Scaled separation *x/h* between a pair of particles as function of the scaled time *t/T* for theory
(continuous line) and an experiment obtained with an oscillating inplane field of magnitude $H_{\|}$ = 14.2 Oe

and a constant normal field $H_{\perp}$ = 0.8$H_{\|}$.

In the current case, the time unit is T = 32 s, and particles come to the predicted
equilibrium distance r = 2.35h after a few minutes. Comparing this data with the present
theory (full line) and with the preexisting expression ignoring the effect of the boundary
conditions along the nonmagnetic plates [4,27] (dashed line) is clear in showing the
importance of the susceptibility contrast for explaining the existence of the secondary
potential minimum.

The dynamical equation ruling the particles in this overdamped regime is obtained
by balancing the Stokes drag from the embedding fluid with the magnetic interactions,
which leads to $\dot{x} = -u'(x)$, where the dot refers to derivation in with respect to the

dimensionless time $t' = t/T$, with $T = 3T_c \cdot h^5 / a^5$ being the unit time. The function
$t'(x)$ was then numerically evaluated as the integral of $-1/u'$ from the initial a/h to
the actual x value of the scaled separation, to obtain the full line of Figure 13.

A range of values of the β parameter was explored in a series of experiments. Some
discrepancy between experiments and theory was found at small particle separation and
is believed to be due to the effect of the non point-like character of the magnetic holes,
which should generate higher order terms in a multipolar expansion at moderate
separations r/a . Qualitatively, the dipolar perturbation field of one hole does not fulfill
properly the boundary conditions along the surface of another close hole, and a repulsive
term corresponding qualitatively to taking into account images of one sphere in the other
one, similar to the repulsive effect of images in the plane boundaries on its source
particle, becomes sensitive at these short distances.

6. Conclusions

In this review we have shown that nonmagnetic particles in magnetized ferrofluids denoted magnetic holes are in many ways ideal model systems to test various forms of particle self assembly and dynamics.

In particular, chaining of magnetic holes show cluster size scaling behavior and for the first time it has been possible to use braid theory and Zipf Relation to describe the dynamics of magnetic holes in ac magnetic fields. Finally, the precise formulation of interactions of magnetic holes in ferrofluid layers has been presented. We have established the effective pair interactions of magnetic holes, submitted to magnetic fields including constant normal components and high frequency oscillating inplane components. Due to the susceptibility contrast along the glass plates, a family of potentials displaying a secondary minimum at finite separation distance has been proven to exist, which should allow trapping of nonmagnetic bodies at tunable distances via the external field.

A system with interactions such as described here, should be relevant for any colloidal suspension of electrically or magnetically polarizable particles constrained in layers. The realization of the detailed effective interaction potentials of this system makes it also a good candidate as an analog model system to study phase transitions, aggregation phenomena in complex fluids, or fracture phenomena in coupled granular/fluid systems.

Acknowledgements

The Research Council of Norway has in part supported the work described here.

References

1. Linton, D. and Walsh, S.T. (2003) Nature Materials, **2**, 289.
2. Wollny, K., Läuger, J., and Huck, S. (2002) Appl. Rheol. **12**, 25.
3. Clausen, S., Helgesen, G., and Skjeltorp, A.T. (1998) Phys. Rev. E **58**, 4229.
4. Skjeltorp, A.T. (1983) Phys. Rev. Lett. **51**, 2306.
5. Helgesen, G., Skjeltorp, A. T., Mors, P. M., Botet, R., and Jullien, R. (1988) Phys. Rev. Lett. **61**, 1736.
6. Promislow, J. H. E., Gast, A. P., and Fermigier, M. J. (1995) Chem. Phys. **102**, 5492.
7. Cernak, J., Macko, P., and Kasparkova, M. (1991) J. Phys. D **24**, 1609.
8. Cernak, J. (1994) J. Magn. Magn. Mater. **132**, 258.
9. Fraden, S., Hurd, A.J., and Meyer, R.B. (1989) Phys. Rev. Lett. **21**, 2373.
10. Meakin, P. (1983) Phys. Rev. Lett. **51**, 1119.
11. Kolb, M., Botet, R., and Jullien, R. (1983) Phys. Rev. Lett. **51**, 1123.
12. Miyazima, S., Meakin, P., and Family, F. (1987) Phys. Rev. A **36**, 1421.

13. Robinson, D. J. and Earnshaw, J. C. (1992) Phys. Rev. A **46**, 2045; **46**, 2055; **46**, 2065.
14. Vicsek, T. and Family, F. (1984) Phys. Rev. Lett. **52**, 1669.
15. Ugelstad J., et al., (1980) Adv. Colloid Int. Sci. **13**, 101.
16. Types EMG905 and EMG909 from Ferrotec Corporation, 40 Simon St., Nashua, NH 03060-3075, USA.
17. Cernak, J., Helgesen, G., and Skjeltorp, A.T. (2004) unpublished.
18. Kristiansen, K. de L., Helgesen, G., and Skjeltorp, A. T. (2004) Physica A **335**, 413.
19. Moore, C. (1993) Phys. Rev. Lett. **70**, 3675.
20. Pieranski, P., et al. (1996) Phys. Rev. Lett. **77**, 1620.
21. Helgesen, G., Pieranski, P., and Skjeltorp, A. T. (1990) Phys. Rev. A **42**, 7271.
22. Elrifai, E. A. and Morton, H. R. (1994) Quart. J. Math. **45**, 479.
23. Zipf, G. K. (1949) *Human Behavior and The Principle of Least Effort* (Addison-Wesley Press, Massachusetts).
24. Mandelbrot, B. (1954) Word **10**, 1.
25. Wang, K.G. and Tokuyama, M. (1999) Physica A **265**, 341.
26. Clausen, S., Helgesen, G., and Skjeltorp, A.T. (1998) Phys. Rev. E **58** 4229.
27. Helgesen, G, and Skjeltorp, A.T. (1991) J. Magn. Magn. Mater. **97**, 25.
28. Toussaint, R., et al. (2004) Phys. Rev. E **69**, 011407.
29. Bleaney, B. and Bleaney, B. (1978) *Electricity and Magnetism* (Oxford University Press).
30. Weber, E. (1950) *Electromagnetic fields: theory and applications*, vol. 1, (Wiley, New York).

Structures in Molecular Networks

K. Sneppen[1,2], S. Maslov[3] and K.A. Eriksen[1]

[1] Nordita, Blegdamsvej 17, 2100 Copenhagen Ø, Denmark
[2]Department of Physics, NTNU, Trondheim, Norway
[3] Department of Physics, Brookhaven National Laboratory, Upton, New York 11973, USA

1   Introduction

Cells are controlled by the action of molecules upon molecules. Receptor proteins in the outer cell membranes, senses the environment and as function of this outer interactions induces changes in the states of specific proteins inside the cell. These protein then again interact and convey the signal further to other proteins, and so forth, until some appropriate action is taken. This states of a protein may e.g. be methylation status, phosphorylation status or allosteric conformation. The final action may be transcription regulation, thereby making more of some kinds of proteins, it may be chemical, or it may be dynamical. A chemical response would be to change the free concentration of a particular protein by binding these to other proteins. A dynamical response could be the activation of some motor, as done in the chemotaxis of *E.coli*.

A particular important action is to change protein composition of the cell. This takes place through genetic regulations. Getice regulation of proteins often take the form illustrated in Fig. 1. Genetic regulation of many proteins, that regulate each other makes a network. Simple genetic networks related to λ phage [1, 2, 3]. and sub parts of the *E.coli* network [4] are well studied, and to some extend modelled in the litterature, see e.g. [5]. The presently known regulatory network of yeast is shown in Fig. 2. The action of proteins in this type of network network is to control the production of proteins.

An interesting common feature of molecular networks is an extremely

A.T. Skjeltorp and A.V. Belushkin (eds.), Forces, Growth and Form in Soft Condensed Matter: At the Interface between Physics and Biology, 181-193.

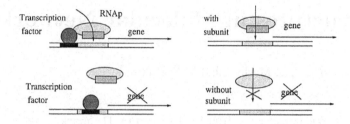

Figure 1: Examples of genetic regulations: on left panel we show respectively positive and negative regulation by a transcription factor (TF). The TF is a protein that binds to a region on the DNA called an operator (dark region on the DNA). The positive regulation is through a binding between the TF and the RNA polymeraze (RNAp), that increases the chance that RNAp binds to the promoter that is shown as medium dark region on the DNA strand. Negative regulation occur when the operator is placed such the bound TF prevents the RNAp from binding to the promoter. On the right panel we show an example of another level of control, associated to presence of sub-units of the RNAp, where certain sub-units allow the RNAp to bind to certain classes of promoters.

Figure 2: Core of regulatory network in yeast, showing all proteins that is known to regulate at least one other protein. Arrows indicate direction of control, that may be either positive or negative. Functionally the network is roughly divided into the upper half that regulate metabolism, and a lower half that regulate cell growth and division. In addition there is a few cell stress response systems in the intersection between these two halves.

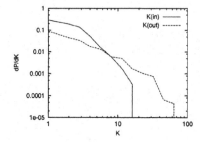 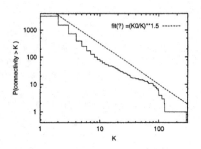

Figure 3: On right panel we show $N(K) = dP/dK$ for regulatory network in yeast, separated into respective the in-link and the out-link connectivity distributions. Notice that out link are broader, not far from $N(K) \propto 1/K^{1.5}$. On left panel we show overall distribution of connectivities in the two hybrid measurement of Ito et al. (2001). We show the cumulative distributions $P(> K)$, as these allows for better judgement of suggested fits. Notice that if $P(> K) \propto 1/K^{1.5}$, then $N(K) = dP/dK \propto 1/K^{2.5}$.

broad, maybe scale-free, distribution of connectivities (defined as the number of immediate neighbors) of their nodes [6]. While the majority of nodes in such networks are each connected to just a handful of neighbors, there exist a few hub nodes that have a disproportionately large number of interaction partners. The histogram of connectivities is an example of a low-level topological property of a network. While it answers the question about how many neighbors a given node has, it gives no information about the identity of those neighbors. It is clear that most of non-trivial properties of networks lie in the exact way their nodes are connected to each other. However, such connectivity patterns are rather difficult to quantify and measure. By just looking at many large complex networks one gets the impression that they are wired in a rather haphazard way. One may wonder which topological properties of a given network are indeed random, and which arose due to evolution and/or fundamental design principles and limitations? Such non-random features can then be used to identify the network and better understand the underlying complex system.

2 A null model for network analysis

In this section we propose a universal recipe for how such information can be extracted. To this end we first construct a proper null randomized model of a given network. An example of such a network is shown in Fig. 2, which indeed have a broad degree distribution as seen in Fig. 3. As was pointed out in [7], broad distributions of connectivities in most real complex networks indicate that the connectivity is an important individual characteristic of a node and as such it should be preserved in any meaningful randomization process. In addition to connectivities one may choose to preserve some other low-level topological properties of the network. Any higher level topological property, such as e.g. the pattern of correlations between connectivities of neighboring nodes, the number of loops of a certain type, the number and sizes of components, the diameter of the network, spectral properties of its adjacency matrix, can then be measured in the real complex network and separately in an ensemble of its randomized counterparts. Dealing with the whole ensemble allows one to put error bars on any quantity measured in the randomized network. One then concentrates only on those topological properties of the complex network that significantly deviate from the null model, and, therefore, are likely to reflect its basic design principles and/or evolutionary history.

The *local rewiring algorithm* that randomizes a network yet strictly conserves connectivities of its nodes [8, 9] consists of repeated application of the elementary rewiring step shown and explained in detail in Fig.4. It is easy to see that the number of neighbors of every node in the network remains unchanged after an elementary step of this randomization procedure. The directed network version of this algorithm separately conserves the number of upstream and downstream neighbors (in- and out-degrees) of every node.

Once an ensemble of randomized versions of a given complex network is generated, the abundance of any topological pattern is compared between the real network and characteristic members of this ensemble. This comparison can be quantified using two natural parameters: 1) the ratio

$$R(j) = \frac{N(j)}{\langle N_r(j) \rangle} \tag{1}$$

where $N(j)$ is the number of times the pattern j is observed in the real network, and $\overline{N_r(j)}$ is the average number of its occurrences in the ensemble of its random counterparts; 2) the Z-score of the deviation defined as $Z(j) =$

186

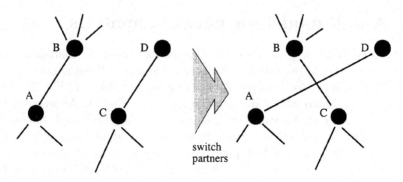

Figure 4: One elementary step of the local rewiring algorithm. A pair of edges A—B and C—D is randomly selected. They are then rewired in such a way that A becomes connected to D, and C - to B, provided that none of these edges already exist in the network, in which case the rewiring step is aborted, and a new pair of edges is selected. The last restriction prevents the appearance of multiple edges connecting the same pair of nodes.

$[N(j) - \overline{N_r(j)}]/\Delta N_r(j)$, where $\Delta N_r(j)$ is the standard deviation of $N_r(j)$ in the randomized ensemble. This general idea was recently applied to protein networks in yeast [8] and E. coli [10], and subsequently applied to analysis of the hardwired Internet (the millennium snapshot of the Internet (data from January 2, 2000), when $N = 6474$ Autonomous Systems were linked by $E = 12572$ bi-directional edge), see Maslov et al [17]. In Fig. 5 we show the qualitative difference between molecular networks analyzed in [8] and the Internet analyzed in [17]. One sees that these networks exhibit roughly opposite hierarchical features. We speculate that this reflects the limited specificity of individual proteins, in distinguishing between different exit channels. In contrast, the Internet is made of computers (autonomous systems), with huge internal specificity.

We stress that the correlation profile is by no means the only topological pattern one can investigate in a given complex network, with other examples being its spectral dimension [13], the betweenness of its edges and nodes [14, 11], feedback circuits, feed-forward loops, and other small network motifs [10]. The interplay between such higher order structures and how they are influenced by the correlation profile of the network is discussed in [17].

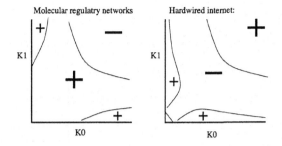

Figure 5: Correlation profile for the hardwired Internet and for regulatory molecular networks in yeast. A + means that one see more of the corresponding connections in the real network than its randomized counterpart, a - implies a relative suppression of the connections. Notice that in protein networks, highly connected nodes tend to avoid connecting to each other. Also notice that medium connected proteins, that is proteins with 3-8 connections, preferentially associate directly to each other. these intermediately connected proteins may form the "computational core of the protein networks.

3 Evolution of Networks

Network are not static. They change in time, and this evolution can in some cases be followed. For molecular networks changes in may occur by gene duplication. In fact, gene duplication followed by functional divergence of associated proteins is a major force shaping molecular networks in living organisms [19]. Thus about 1/3 of the proteins in most organism have gene duplicates within the same organism. A pair of proteins which are generated by such a gene duplication is called paralogs. Recent availability of system-wide data for yeast S. Cerevisiae [20, 21, 22, 23] have allowed us [18] to access the effects of gene duplication on robustness and plasticity of molecular networks.

To this end we have measured (see Fig. 8: 1) The similarity of positions of duplicated genes in the transcription regulatory network [20] given by the number of transcription regulators they have in common (their upstream regulatory overlap); 2) The similarity of the set of binding partners [21, 22] of their protein products (their downstream overlay), and 3) their ability to substitute for each other in knock-out experiments [23]. These measures reflect, correspondingly, the upstream and downstream properties of molecular

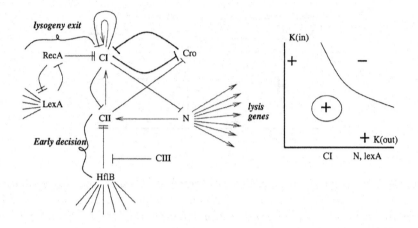

Figure 6: Lambda phage network with highly (like LexA and N), respectively lowly connected proteins (like CI, CII and Cro) in the network. The actual computation of the phage is performed through moderately connected CI, Cro and CII, whereas the overall input and output is directed through the highly connected proteins N and LexA (and protease HflB from E.coli). On right side of figure we show the correlation profile, indicating the relative contributions of the lambda proteins. Notice that a line with one vertical end line indicates negative regulation. A line with two vertical end lines indicate that active degradation through protease activity can take place.

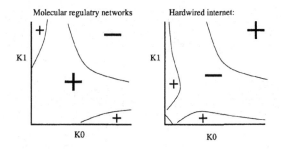

Figure 7: Illustration of the concept of overlap in a molecular network. For a pair of paralogs the overlap Ω is defined as the number of common neighbors they have in the network. In the case of transcription network the regulatory overlap Ω_{reg} counts transcription factors regulating both paralogs, while for the physical interaction network the interaction overlap Ω_{int} counts their common binding partners. The pair of paralogs used in this illustration has the overlap $\Omega = 2$ out of the total of 5 distinct neighbors of the pair. That corresponds to a normalized overlap of $2/5 = 0.40$.

networks around the duplicated genes.

In any case a time development of the properties 1-3) is estimated by averaging over a huge number of protein pairs which all have similar divergence in sequence from each other. When the proteins are newly duplicated they are nearly identical, when they are maybe 80% identical they have duplicated for maybe 100 million years ago, and when they are even weaker related to each other the time since divergence is even longer. In any case the identification of time with relative similarity is very uncertain, and in the analysis we therefore only use similarity as a measure in itself, and thus compare divergence of functions around a protein pair, with divergence of its intrinsic sequences.

We found [18] that the upstream transcriptional regulation of duplicated genes diverges fast, losing on average 4% of their common transcription factors for every 1% divergence of their amino acid sequences. In contrast, the set of physical interaction partners of their protein products changes much slower. The relative stability of downstream functions of duplicated genes, is further corroborated by their ability to substitute for each other in gene knockout experiments. We believe that the combination of the upstream

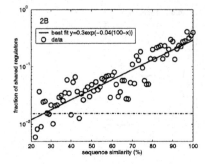 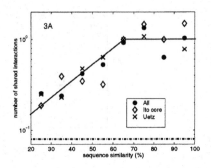

Figure 8: On left panel we show decay of overlap of upstream transcription regulation, with decreasing similarity of proteins, from [18]. As times since duplication increases, similarity decreases, and similarity in who regulates the two proteins decreases even faster. On right panel we show the decay in downstream overlap between proteins that physically bind to the two paralogs.

plasticity and the downstream robustness is a general feature determining the evolvability of molecular networks.

Analysis of these types of network evolution data are also found in [24, 25, 26].

4 Conclusions

In summary we have proposed a general algorithm to detect characteristic topological features in a given complex network. In particular, we introduced the concept of the *correlation profile*, which allowed us to quantify differences between different complex networks even when their connectivity distributions are similar to each other. Applied to the Internet, this profile identifies hierarchical features of its structure, and helps to account for the level of clustering in this network.

Finally we briefly outlined the possibility for analyzing evolution of molecular networks, by utilizing genome as a pale-ontological record, and focusing in particular on the about 1/3 of the genes in any organism that have duplicates within the same organism. This have lead us to the general conclusion that molecular networks use downstream robustness to facilitate fast upstream evolvability of network rewiring. This allow living organisms to use old proteins in new situations, and thereby develop new traits and possibly new biological species. Future studies may teaches us about interplay between evolution and design principles of working molecular networks, see [27].

Acknowledgement(s)

Work at Brookhaven National Laboratory was carried out under Contract No. DE-AC02-98CH10886, Division of Material Science, U.S. Department of Energy.

References

[1] G. Ackers, A. Johnson & M. Shea (1982). *Quantitative model for gene regulation by λ phage repressor. Proc. Natl. Acad. Sci. USA* **79**, 1129-1133.

[2] H.H. McAdams & A. Arkin (1997). *Stochastic mechanisms in gene expression. Proc. Natl. Acad. Sci. U S A* **94**, 814-819.

[3] E. Aurell, S. Brown and K. Sneppen (2002). *Stability puzzles in phage* λ, Phys. Rev. E (2002).

[4] S.S. Shen-Orr, R. Milo, S. Mangan and U. Alon (2002). *N*etwork motifs in the transcriptional regulation of Escherichia coli Nature Genetics, Published online: 22 April 2002, DOI:10. 1038/ng88

[5] K.B. Arnvig, S. Pedersen and K. Sneppen (2000). "*Thermodynamics of Heat Shock Response*". Phys. Rev. Lett. **84** 3005 (cond-mat/9912402)

[6] H. Jeong, B. Tombor, R. Albert, Z. N. Oltvai, A.-L. Barabasi, Nature, **407**, 651–654 (2000). H. Jeong, S. Mason, A.-L. Barabasi, Z.N. Oltvai, Nature **411**, 41–42 (2001).

[7] M. E. J. Newman, S. H. Strogatz, and D. J. Watts, Phys. Rev. E, **64**, 026118 ,1-17 (2001); M. E. J. Newman, cond-mat/0202208. to appear

in Handbook of Graphs and Networks, S. Bornholdt and H. G. Schuster (eds.),

[8] S. Maslov and K. Sneppen, Science, **296**, 910 (2002).

[9] These algorithms first appeared in the context of random matrices in: D. Gale, Pacific J. Math., **7**, 1073-1082 (1957); H.J. Ryser, in *"Recent Advances in Matrix Theory"*, pp. 103-124, Univ. of Wisconsin Press, Madison, (1964). More recently they were used in the graph-theoretical context: R. Kannan, P. Tetali, S. Vempala, Random Structures and Algorithms **14**, 293-308, (1999).

[10] S. S. Shen-Orr, R. Milo, S. Mangan, and U. Alon, Nature Genetics, **31**,64 (2002).

[11] R. Pastor-Satorras, A. Vazquez, and A. Vespignani, Phys. Rev. Lett. **87**, 258701 1-4 (2001);

[12] Website maintained by the NLANR Measurement and Network Analysis Group at http://moat.nlanr.net/

[13] S. Bilke and C. Peterson, Phys. Rev. E **64**, 036106 (2001).

[14] M. Girvan and M. E. J. Newman, cond-mat/0112110 (2001).

[15] D. Watts and S. Strogatz, Nature **293**, 400 (1998).

[16] N. Metropolis, *et al.*, J. Chem. Phys. **21**, 1087 (1953).

[17] S. Maslov, K. Sneppen, A. Zaliznyak, To appear in PNAS, cond-mat/0205379 (2002). Submitted to PNAS.

[18] S. Maslov, K. Sneppen, K.A. Eriksen, Submitted to Trends in Genetics. 2003.

[19] S. Ohno. 1970. *Evolution by gene duplication.* Springer-Werlag, Berlin.

[20] T.I. Lee, *et al.* 2002. *Science* **298**, 799-804.

[21] P. Uetz et al. (2000). *Nature* **403**: 623-627.

[22] T. Ito et al. (2002)' *Proc. Natl. Acad. Sci. USA* **98**:, 4569-4574.

[23] G. Giaever *et al.* *Nature* **418**, 387-391.

[24] Z. Gu et al. (2002) *Trends in Genetics* **18:** 609-613.

[25] A. Wagner. *Mol. Biol. Evol.* **18:** 1283-1292.

[26] Z. Gu *et al.* *Nature* **421**, 63-66 (2003)

[27] S. Bornholdt and K.Sneppen (2000). Proc Roy. Soc. London, B **267** 2281.

Oscillating Gene Expressions in Regulatory Networks.

M. H. Jensen[1], K. Sneppen[1,2] and G. Tiana[1,3]

[1] The Niels Bohr Institute, Blegdamsvej 17, 2100 Copenhagen, Denmark

[2] NORDITA, Blegdamsvej 17, 2100 Copenhagen, Denmark

[3] Dept. of Physics, University of Milano, via Celoria 16, 20133 Milano, Italy

Abstract

We review our work on oscillating genetic expressions in two regulatory biological networks. The first is the Hes1 protein/hes1 mRNA feedback system which regulates timing in many cell types. The second is the p53-mdm2 protein regulatory system which is important for cell control after DNA damage. In both cases, we formulate sub networks of the full genetic network, by identifying the key players in the genetic regulation. We formulate in each case differential equations of two variables and introduce, based on the biological properties, a time-delay in the equations in order to generate the oscillatory behavior [1, 2]. The obtained results are in very good agreement with experimental measurements.

1 The Hes1 regulatory protein system.

A number of genes change their expression pattern dynamically by displaying oscillations. In a few important cases these oscillations are sustained and can work as molecular clocks, as in the well known cases of the circadian clock [3] and the cell cycle [4]. In other cases the oscillations in protein expression are connected with the response to external stimuli, as reported for protein p53 after induction by DNA damage [5] or as reported in association to specificity in gene expression [6]. Recently oscillations have been observed for the Hes1 system studied in Ref. [7]. The Hes1 system is particularly interesting because it is connected with cell differentiation, and the temporal oscillations of the Hes1 system may thus be associated with the formation of spatial patterns in development. Oscillations can be obtained by a closed loop of inhibitory couplings, provided that there are at least 3 different elements

A.T. Skjeltorp and A.V. Belushkin (eds.), Forces, Growth and Form in Soft Condensed Matter: At the Interface between Physics and Biology, 195-202.

[7, 8]. However, we noted in the study of the p53 network [2] that a time delay in one of the components can give rise to oscillations also in a system composed of only two species.

In the Hes1 system, the protein Hes1 represses the transcription of its own mRNA, and the system displays oscillations in both the concentration of the protein and of its mRNA. We want to test the hypothesis that Hes1 and its mRNA are sufficient ingredients to produce oscillations in the system and propose that this system can be modelled by a two-dimensional dimensionless differential equation in terms of the concentrations $[mRNA]$ and $[Hes1]$ ([1]):

$$\frac{d[mRNA]}{dt} = \frac{\alpha k^h}{k^h + [Hes1(t-\tau)]^h} - \frac{[mRNA(t)]}{\tau_{rna}} \qquad (1)$$
$$\frac{d[Hes1]}{dt} = [mRNA(t)] - \frac{[Hes1(t)]}{\tau_{hes1}}$$

The meaning of these equations is that mRNA is produced at rate α when Hes1 is bound to the DNA. The probability that Hes1 is bound to DNA is $k^h/(k^h + [Hes1]^h)$, where k is a characteristic concentration for dissociation of Hes1 from the DNA, and h is the Hill coefficient that takes into account the cooperative character of the binding process. Moreover, Eqs. (1) say that mRNA undergoes degradation with characteristic time τ_{rna}, that the production rate of Hes1 is proportional to the concentration of mRNA and that Hes1 is degraded on the time scale τ_{hes1}. Note that the terms associated with degradation in Eqs. (1) not only describe the spontaneous degradation of the protein, but also the outflow caused by the protein going to interact with other parts of the cell.

The key point is that the production of mRNA is delayed by a time τ, which takes into account the lengthy molecular processes involved in the system (translation, transcription, etc.). If one inserts the delay in the production of Hes1 (the second of Eqs. 1), instead that of mRNA, the results remains very similar to the ones reported here.

An important factor which determines the cooperativity in the production of mRNA is the fact that Hes1 is a dimer, and consequently we expect that the Hill coefficient h is of the order of 2. On the other hand, its precise value is not known. We have repeated our calculations for different values of h (i.e., $h = 1.5, 2, 4$) and found that the system displays oscillations in all cases analyzed, although the detailed features of these oscillations (e.g., those displayed in Table 1) depend on the particular choice of h. This result

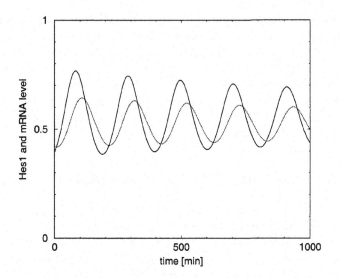

Figure 1: The oscillatory behavior of the concentration $[Hes1]$ of the protein Hes1 (dashed curve) and mRNA (solid curve), as calculated from Eq. (1). The following parameters are used: $\tau_{rna} = 24.1$ min, $\tau_{hes1} = 22.3$ min, $\alpha = 1 \ [R]_0/$min, $\beta = 0.1$ min^{-1}, $k = 0.1[R]_0$, $h = 2$, $\tau = 50$ min, and the plot show concentrations in units of $[R]_0$ (from [1]).

agrees with the fact that the physical reason which causes oscillations is not the nonlinearity of the equations but the delay. In the following we analyze in detail the case $h = 2$.

From Ref. [7], τ_{rna} and τ_{hes1} are of the order of 25 minutes. The value of the time delay is difficult to assess, since it is determined by a variety of molecular processes. One can guess that its order of magnitude is tens of minutes.

The solution of Eqs. (1) is displayed in Fig. 1. For the chosen set of parameters, the system displays damped oscillations with period $\Delta\tau \approx 170$ min and damping time $\tau_{damp} \sim 9500$ min. The dependence of $\Delta\tau$ and τ_{damp} on the delay τ is listed in Table 1. The oscillation period stays constant for low value of the delay and increases as $\tau \gg \tau_{rna}$. Also the damping time increases with τ, the oscillations becoming sustained for $\tau > 80$ [1].

For any delay in the range $10 < \tau < 50$ min, the oscillation period is consistent with that found experimentally, and also the time difference between the peaks in Hes1 and mRNA is 18 min, similar to the experimental findings. For $\tau < 10$ min, the system shows no oscillations. To check the robustness of the results, we have varied α, β and k over 5 orders of magnitude around the basal values listed in the caption to Fig. 1, and observed no qualitative difference with the oscillatory behaviour described above. On the other hand, an increase of τ_{hes1} and τ_{rna} disrupts the oscillatory mechanism. This is because these two quantities set the time scale of the system, with which τ has to be compared.

2 The p53-mdm2 regulatory protein systems.

In healthy cells, a loopback mechanism involving the protein p53 is believed to cause growth arrest and apoptosis as a response to DNA damage [9, 10, 11, 12]. Under normal conditions the amount of p53 protein in the cell is kept low by a genetic network built of the mdm2 gene, the mdm2 protein and the p53 protein itself. p53 is produced at a essentially constant rate and promotes the expression of the mdm2 gene [5]. On the other hand, the mdm2 protein binds to p53 and promotes its degradation [13], decreasing its concentration. When DNA is damaged, a cascade of events causes phosphorylation of several serines in the p53 protein, which modifies its binding properties to mdm2 [14]. As a consequence, the cell experiences a sudden increase in the concentration of p53, which activates a group of genes, responsible for cell cycle arrest and apoptosis. This increase in p53 can reach values of the order of 16 times the basal concentration [15].

A qualitative study of the time dependence of the concentration of p53 and mdm2 has been carried out in ref. [5]. Approximately one hour after the stress event (i.e., the DNA damage which causes phosphorylation of p53 serines), a peak in the concentration of p53 is observed, lasting for about one hour. This peak partially overlaps with the peak in the concentration of mdm2, lasting from ≈ 1.5 to ≈ 2.5 hours after the stress event. Another small peak in the concentration of p53 is observed after several hours.

The model we suggest for this network is described in Fig. 2 ([2]). The total number of p53 molecules, produced at constant rate S, is indicated with p. The amount of the complexes built of p53 bound to mdm2 is called pm. These complexes cause the degradation of p53 (through the ubiquitin pathway), at a rate a, while mdm2 re–enters the loop. Furthermore, p53 has a spontaneous decay rate b. The total number of mdm2 proteins is indicated as

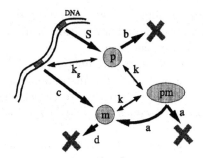

Figure 2: A sketch of the loopback mechanism which control the amount of p53 in the cell. The grey crosses indicate that the associated molecule leaves the system (from [2]).

m. Since p53 activates the expression of the mdm2 gene, the production rate of mdm2 is proportional (with constant c) to the probability that the complex p53/mdm–gene is built. We assume that the complex p53/mdm2–gene is at equilibrium with its components, where k_g is the dissociation constant and only free p53 molecules (whose amount is $p - pm$) can participate into the complex. The protein mdm2 has a decay rate d. The constants b and d describe not only the spontaneous degradation of the proteins, but also their binding to some other part of the cell, not described explicitly by the model. The free proteins p53 and mdm2 are considered to be at equilibrium with their bound complex pm, and the equilibrium constant is called k.

The dynamics of the system can be described by the equations

$$\frac{\partial p}{\partial t} = S - a \cdot pm - b \cdot p \tag{2}$$

$$\frac{\partial m}{\partial t} = c\frac{p(t - \tau) - pm(t - \tau)}{k_g + p(t - \tau) - pm(t - \tau)} - d \cdot m$$

$$pm = \frac{1}{2}\left((p + m + k) - \sqrt{(p + m + k)^2 - 4p \cdot m}\right).$$

In the second equation we allow a delay τ in the production of mdm2, due to the fact that the transcription and translation of mdm2 lasts for some time after that p53 has bound to the gene.

For the choice of the numerical parameters, see [2]. In the case that the production of mdm2 can be regarded as instantaneous (no delay, $\tau = 0$), the concentration of p53 is rather insensitive to the change of the dissociation constant k. The stationary values of p and m are found as fixed points of the Eq. 2 [2].

The dynamics changes qualitatively if we introduce a nonzero delay in Eqs. 2. Keeping that the halflife of an RNA molecule is of the order of 1200 s [16], we repeat the calculations with $\tau = 1200$. The Eqs. 2 are solved numerically, starting from the conditions $p(0) = 0$ and $m(0) = 0$ and making use of a variable–step Adams algorithm. After the system has reached its stationary state under basal condition, a stress is introduced (at time $t = 20000$ s) by changing instantaneously the dissociation constant k. In Fig. 3 we display a case in which the stress multiplies k by a factor 15 (a), a case in which it divides it by a factor 15 (b) and by a factor 5 (c). When k is increased by any factor, the response is very similar to the response of the system without delay (cf. e.g. Fig. 3a). On the contrary, when k is decreased the system displays an oscillatory behaviour. The maximum of the first peak takes place approximately 1200s after the stress, which is consistent with the lag–time observed in the experiment [5], and the peaks are separated from ≈ 2300s.

The minimum value of the delay which gives rise to the oscillatory behaviour is $\tau \approx 100$s. For values of the delay larger than this threshold, the amplitude of the response is linear with τ. The lag time before the p53 response is around 3000s (in accordance with the 1h delay observed experimentally [5]) and is independent on all parameters, except c and τ. The dependence of the lag time on τ is approximately linear up to 5000s (the longest delay analyzed). Increasing c the lag time increases to 8000s (for $c = 10^4$) and 25000s (for $c = 10^5$). On the other hand, the period of oscillation depends only on the delay τ, being approximately linear with it. The minimum value of the delay which gives rise to the oscillatory behaviour is $\tau \approx 100$s. For values of the delay larger than this threshold, the amplitude of the response is linear with τ.

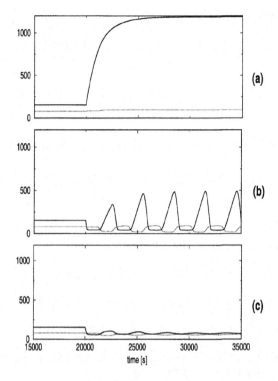

Figure 3: The response in the concentration of p53 (solid line) and mdm2 (dotted line) upon variation of the dissociation constant k. At time 20000 s the constant k is multiplied by 15 (a), divided by 15 (b) and divided by 5 (c) (from [2]).

References

[1] M.H. Jensen, K. Sneppen and G. Tiana, FEBS Letters **541**, 176-177 (2003).

[2] G. Tiana, M. H. Jensen, K. Sneppen, Europ. Phys. J. B **29**, 135 (2002).

[3] M.P. Antoch, et al. *Cell*, **89**: 655-67 (1997).

[4] B. Novak & J.J. Tyson, J. Theor. Biol. **173**, 283-305 (1995).

[5] Y. Haupt, R. Maya, A. Kazaz and M. Oren, Nature **387** 296 (1997).

[6] A. Hoffmann, A. Levchenko, M.L. Scott & D. Baltimore, Science **298** 1241-1245 (2002).

[7] H. Hirata et al. *Science* **298** 840 (2002).

[8] M.B. Elowitz, S. Leibler, *Nature* **403**, 335 (2002).

[9] B. Vogelstein, D. Lane and A. J. Levine, Nature **408** (2000) 307–310

[10] M. D. Shair, Chemistry and Biology **4** (1997) 791–794

[11] M. L Agarwal, W. R. Taylor, M. V. Chernov, O. B. Chernova and G. R. Stark, J. Biol. Chem. **273** (1998) 1–4

[12] R. L. Bar–Or *et al.*, Proc. Natl. Acad. Sci. USA **97** (2000) 11250

[13] M. H. G. Kubbutat, S. N. Jones and K. H. Vousden, Nature **387** (1997) 299–393

[14] T. Unger, T. Juven–Gershon, E. Moallem, M. Berger, R. Vogt Sionov, G. Lozano, M. Oren and Y. Haupt, EMBO Journal **18** (1999) 1805–1814

[15] J. D. Oliner, J. A. Pietenpol, S. Thiagalingam, J. Gyuris, K. W. Kinzler and B. Vogelstein, Nature **362** (1993) 857–860

[16] F. S. Holstege *et al.*, Cell **95**, 717 (1995)

TRANSPORT PROPERTIES OF SEGMENTED POLYMERS AND NON-SPHERICAL NANOPARTICLES STUDIED BY BROWNIAN DYNAMICS SIMULATIONS

S.N. NAESS and A. ELGSAETER
Department of Physics, Norwegian University of Science and Technology (NTNU), Høgskoleringen 5, N-7491 Trondheim, Norway

Abstract. We have succeeded in extending the Brownian dynamics analyses to systems consisting of non-spherical nanoparticles interconnected by conservative forces or holonomic constraints. The formal theory takes fully into account both nanoparticle size and surface topography. Our theory also incorporates stretching, bending and torsional stiffness between nearest neighbor subunits, excluded volume effects, external force fields, fluid flow and fluiddynamic interactions. The generalized conformation-space diffusion equations are rigorously derived from kinetic theory. The equivalent stochastic differential equation are used as our basis for development of the associated Brownian dynamics algorithms. These algorithms may be employed to carry out equilibrium as well as non-equilibrium simulations of the conformational dynamics and transport properties for a wide class of nanoparticle systems including segmented polymers.

1. Introduction

Many biopolymers with secondary and higher order structure consist of subunits that can move relative to one another. Depending on the stretching, bending and torsional stiffness of the joints interconnecting the subunits, the polymers may undergo large-scale, thermally induced conformational changes that are stochastic and often play a functional role. The stochastic conformational dynamics plays a similarly important role in liquid crystals made up of e.g. inorganic non-spherical nanoparticles. Effects caused by these large-scale conformational fluctuations can most often only be studied by Brownian dynamics simulations. This means using a nanometer length scale and nanosecond time scale. The viscous fluid (solvent) in which the biopolymers are embedded, is modeled as a continuum.

At the nanometer length scale biopolymers in general display irregular surface topography. Also the subunit mass distribution is usually non-

A.T. Skjeltorp and A.V. Belushkin (eds.), Forces, Growth and Form in Soft Condensed Matter: At the Interface between Physics and Biology, 203-216.

symmetric. For most biopolymers this makes it mandatory to include all three degrees of rotational freedom for each subunit in order to achieve a full description of the dynamics. To carry out detailed analyses of the fluiddynamic forces it is necessary to incorporate the detailed surface topography of each subunit. The bead-spring and bead-rod polymer models comprise only structureless subunits and are therefore in general poorly suited for quantitative analysis of biopolymer dynamics. Brownian dynamics simulation of biopolymers consisting of rigid subunits may in principle be carried out using the standard numerical algorithms for bead-spring polymer chains [1] provided that the rigid subunits with arbitrary shape are modeled as a sufficiently large number of densely packed structureless beads interconnected by stiff springs [2]. Efforts have also been made to substitute all or some of these stiff springs with holonomic constraints [3]. One major drawback of these methods is that all the beads must be incorporated explicitly in the dynamics simulations. This is necessary despite that the detailed dynamics of these beads has no physical significance. In addition the accuracy is limited when the fluiddynamic equations describing fluid flow around single spheres are used as the basis for modeling of the fluid flow near densely packed assemblies of spheres. As a consequence only the simplest cases can be studied successfully using this approach.

History has shown that, except for the simplest cases, basing nanoparticle and biopolymer dynamics analyses on stochastic equations not founded rigorously on kinetic theory can be treacherous. The key pioneering contributions to the kinetic theory of polymers in solution appeared more than half a century ago in the seminal work by Kramers [4]. During the years that followed Kirkwood [5], Rouse [6], Zimm [7] and others made major contributions to the theory for polymers consisting of structureless subunits. Important early contributions to the understanding of the relation between kinetic and stochastic theory for these coarse-grained polymer models were made by Zwanzig [8] and Fixman [9]. Comprehensive reviews of kinetic and stochastic theory for polymer systems containing structureless subunits can be found in the monographs by Bird et al. [10] and Öttinger [11], respectively.

Using kinetic theory as our starting point we report on the generalized conformation-space diffusion (Fokker-Planck) equations, the associated Itô stochastic differential equations and the numerical Brownian dynamics algorithms for polymers consisting of non-spherical subunits interconnected by conservative (spring-like) forces or holonomic constraints (ball-socket joints). We denote these polymer models nugget-spring chains and nugget chains, respectively. The numerical algorithms may carry out equilibrium as well as non-equilibrium simulations. Our formal analysis incorporates stretching, bending and torsional stiffness between nearest neighbor subunits, excluded volume effects, external force fields, fluid flow and fluid-dynamic interactions. Although the inspiration for the present work was applications to challenges associated with biopolymers, the described formalism is applicable to practically every molecular and nanoparticle system amenable to Brownian dynamics including liquid crystals.

2. Microfluiddynamics

The relation between the deterministic fluiddynamic forces acting on a polymer and the resulting deterministic subunit velocities plays a crucial role in the description of both the deterministic and stochastic dynamics of polymer chains in solution when the solvent is modeled as a continuum.

We will in the following use the spatial vector $\vec{r}_\mu^*$ to describe both the Cartesian position and the angular orientation of polymer nugget μ

$$\vec{r}_\mu^* := \{\vec{r}_\mu^{(\text{trans})}, \vec{r}_\mu^{(\text{rot})}\}. \tag{1}$$

Superscript $*$ is throughout this work used to denote that the parameter in question has components which describe both Cartesian position and angular orientation. Vector $\vec{r}_\mu^{(\text{trans})}$ is here chosen to be the center of mass position vector of nugget μ in the Cartesian laboratory coordinate system. The three components of $\vec{r}_\mu^{(\text{rot})}$ give the angular orientation of nugget μ. These three components may be the Eulerian angles [12] ϕ, θ and ψ. Alternatively $\vec{r}_\mu^{(\text{rot})}$ may be the Cartesian rotation vector [13] $\vec{a}(t) = \{a_1(t), a_2(t), a_3(t)\} := \Phi(t)\vec{\delta}^{(a)}(t)$, where Φ is the angle of rotation appearing in the Euler theorem [12] and $\vec{\delta}^{(a)}$ is the unit vector of $\vec{a}$.

When only deterministic dynamics, small Reynolds numbers and nugget-spring polymer chains consisting of N subunits embedded in a Newtonian fluid are considered the nugget spatial velocity reads ($\mu = 1, 2, \ldots, N$)

$$\begin{pmatrix} \dot{\vec{r}}_\mu^{(\text{trans})} \\ \dot{\vec{r}}_\mu^{(\text{rot})} \end{pmatrix} - \begin{pmatrix} \vec{v}^{(\text{trans},\kappa)}(\vec{r}_\mu^*) \\ \vec{v}^{(\text{rot},\kappa)}(\vec{r}_\mu^*) \end{pmatrix} :=$$

$$-\sum_{\nu=1}^{N} \begin{pmatrix} \overset{\Rightarrow(\text{TT})}{\mu}_{\mu\nu} & \overset{\Rightarrow(\text{TR})}{\mu}_{\mu\nu} \\ \overset{\Rightarrow(\text{RT})}{\mu}_{\mu\nu} & \overset{\Rightarrow(\text{RR})}{\mu}_{\mu\nu} \end{pmatrix}^{(\text{ns})} \cdot \begin{pmatrix} \vec{F}_\nu^{(\text{trans},\text{fd})} \\ \vec{F}_\nu^{(\text{rot},\text{fd})} \end{pmatrix}, \tag{2}$$

where superscript $^{(\text{ns})}$ denotes that the parameter in question is associated with nugget-spring chains. This may be rewritten as

$$\dot{\vec{r}}_\mu^* - \vec{v}^{*(\kappa)}(\vec{r}_\mu^*) := -\sum_{\nu=1}^{N} \overset{\Rightarrow*(\text{ns})}{\mu}_{\mu\nu} \cdot \vec{F}_\nu^{*(\text{fd})}, \tag{3}$$

where $\vec{v}^{*(\kappa)}(\vec{r}_\mu^*)$ equals the fluid velocity field at the spatial position $\vec{r}_\mu^*$. In general, this velocity field contains non–zero rotational components. The vector $\vec{F}_\nu^{*(fd)}$ is the deterministic part of the spatial fluiddynamic force from the fluid on nugget ν. The grand spatial mobility tensor for the nugget-spring polymer chain as a whole is denoted $(\overset{\Rightarrow *(ns)}{\boldsymbol{\mu}_{\mu\nu}})$. The 6×6 spatial mobility tensor $\overset{\Rightarrow *(ns)}{\boldsymbol{\mu}_{\mu\mu}}$ describes the microscopic mobility of free individual nuggets. When $\mu \neq \nu$ the components of the grand spatial mobility tensor describe the fluiddynamic interaction between nugget μ and ν. The components of $\overset{\Rightarrow *(ns)}{\boldsymbol{\mu}_{\mu\nu}}$ account for translational-translational (superscript $^{(TT)}$), translational-rotational (superscript $^{(TR)}$), rotational-translational (superscript $^{(RT)}$) and rotational-rotational (superscript $^{(RR)}$) fluiddynamic interaction. The mobility tensor is symmetric and positive semidefinite.

When only deterministic dynamics is studied and the inertia forces on the nuggets can be neglected the spatial force balance for each nugget reads $(\nu = 1, 2, \ldots, N)$

$$\vec{F}_\nu^{*(fd)} + \vec{F}_\nu^{*(\Phi)} + \vec{F}_\nu^{*(e)} = \vec{0}^*, \tag{4}$$

where $\vec{F}_\nu^{*(\Phi)}$ is the conservative spatial force derived from intra-chain potentials (including excluded volume forces) and $\vec{F}_\nu^{*(e)}$ is the conservative spatial force due to external potentials (including the nugget buoyancy). The nugget surface topography affects the polymer chain dynamics partly through the grand spatial mobility tensor and partly through the excluded volume interactions.

The presence of ball-socket joints between all nearest neighbor subunits in a nugget chain can be incorporated in the formal description by including $N - 1$ time–independent and holonomic constraint conditions $(j = 1, 2, \ldots, N - 1)$

$$\vec{g}_j = \vec{0}, \tag{5}$$

where the constraint functions are

$$\vec{g}_j(\vec{r}_{\mu+1}^{(-)} - \vec{r}_\mu^{(+)}) := \vec{r}_{\mu+1}^{(-)} - \vec{r}_\mu^{(+)}. \tag{6}$$

The Cartesian vectors $\vec{r}_\mu^{(+)}$ and $\vec{r}_{\mu+1}^{(-)}$ are the position vectors in the laboratory coordinate system for junction $j = \mu$ (or points of spring attachment in the case of nugget-spring chains) on the surface of nugget μ and $\mu + 1$, respectively.

We have earlier [14, 15] shown that the grand spatial velocity of nugget chains reads

$$
\begin{aligned}
\dot{\vec{r}}^{*(\mathrm{n})} &= -\overset{\Rightarrow}{\boldsymbol{P}}{}^{*} \cdot \overset{\Rightarrow}{\boldsymbol{\mu}}{}^{*(\mathrm{ns})} \cdot [\vec{\boldsymbol{F}}^{*(\mathrm{fd})} - \vec{\boldsymbol{F}}^{*(\kappa)}] \\
&:= \overset{\Rightarrow}{\boldsymbol{\mu}}{}^{*(\mathrm{n})} \cdot [\vec{\boldsymbol{F}}^{*(\Phi)} + \vec{\boldsymbol{F}}^{*(\mathrm{e})} + \vec{\boldsymbol{F}}^{*(\kappa)}],
\end{aligned}
\tag{7}
$$

where superscript $^{(\mathrm{n})}$ denotes that the parameter in question is associated with nugget chains. The spatial projection operator can be expressed as

$$
\overset{\Rightarrow}{\boldsymbol{P}}{}^{*} = \overset{\Rightarrow}{\boldsymbol{\delta}}{}^{*} - \overset{\Rightarrow}{\boldsymbol{\mu}}{}^{*(\mathrm{ns})} \cdot \left(\frac{\partial \vec{g}}{\partial \vec{r}^{*}}\right) \cdot \left[\left(\frac{\partial \vec{g}}{\partial \vec{r}^{*}}\right)^{T} \cdot \overset{\Rightarrow}{\boldsymbol{\mu}}{}^{*(\mathrm{ns})} \cdot \left(\frac{\partial \vec{g}}{\partial \vec{r}^{*}}\right)\right]^{-1} \cdot \left(\frac{\partial \vec{g}}{\partial \vec{r}^{*}}\right)^{T}. \tag{8}
$$

Throughout this text we have adopted the following more compact notation for dyads $\left(\frac{\partial \vec{g}}{\partial \vec{r}^{*}}\right) := \left(\frac{\partial}{\partial \vec{r}^{*}} \vec{g}\right)$. Note that $\vec{\boldsymbol{F}}^{*(\Phi)}$ does not include the constraint forces and $\vec{\boldsymbol{F}}^{*(\kappa)}$ is an equivalent force given by $\vec{\boldsymbol{v}}^{*(\kappa)} := \overset{\Rightarrow}{\boldsymbol{\mu}}{}^{*(\mathrm{n})} \cdot \vec{\boldsymbol{F}}^{*(\kappa)}$.

In the simplest nugget-spring chains the spring potential energy depends only on the spring lengths. The length of these Hookean springs is given by Eq. (6). In the limit of infinitely large Hookean spring constants ($H^{(\mathrm{s})} \to \infty$) the length of each spring equals zero, which means that for this limit we get the same requirement as given in Eq. (5) for the analysis of holonomic constraints. The same mathematical steps used to calculate the mobility tensor for nugget chains can therefore be used to calculate the mobility tensor for nugget-spring chains when $H^{(\mathrm{s})} \to \infty$. The only difference is that in the latter case we calculate spring rather than constraint forces. In the limit $H^{(\mathrm{s})} \to \infty$ we get that the grand spatial velocity of nugget-spring chains equals

$$
\begin{aligned}
\dot{\vec{r}}^{*(\mathrm{ns},\infty)} &= -\overset{\Rightarrow}{\boldsymbol{P}}{}^{*} \cdot \overset{\Rightarrow}{\boldsymbol{\mu}}{}^{*(\mathrm{ns})} \cdot [\vec{\boldsymbol{F}}^{*(\mathrm{fd})} - \vec{\boldsymbol{F}}^{*(\kappa)}] \\
&= \overset{\Rightarrow}{\boldsymbol{\mu}}{}^{*(\mathrm{ns},\infty)} \cdot [\vec{\boldsymbol{F}}^{*(\Phi)} + \vec{\boldsymbol{F}}^{*(\mathrm{e})} + \vec{\boldsymbol{F}}^{*(\kappa)}] \\
&= \overset{\Rightarrow}{\boldsymbol{\mu}}{}^{*(\mathrm{n})} \cdot [\vec{\boldsymbol{F}}^{*(\Phi)} + \vec{\boldsymbol{F}}^{*(\mathrm{e})} + \vec{\boldsymbol{F}}^{*(\kappa)}],
\end{aligned}
\tag{9}
$$

where superscript $^{(\mathrm{ns},\infty)}$ denotes that the parameter in question is associated with a nugget-spring chain in the limit $H^{(\mathrm{s})} \to \infty$. In the latter equation the spatial force $\vec{\boldsymbol{F}}^{*(\Phi)}$ does not include the spring force, just as $\vec{\boldsymbol{F}}^{*(\Phi)}$ in Eq. (7) does not include the constraint force. This means that when the ball-socket joints of a nugget chain embedded in a fluid are substituted by infinitely stiff Hookean springs the deterministic dynamics expressed in terms of spatial coordinates remains unchanged, but the numbers of degrees of freedom for these two cases are different. However, for large, but finite $H^{(s)}$ the deterministic dynamics of a nugget-spring chain may be experimentally indistinguishable from what is described by Eq. (9).

The spatial coordinates constitute a set of generalized coordinates for the nugget-spring chains, but this is not the case for the nugget chains despite that the deterministic dynamics presented in Eq. (7) is given in terms of spatial coordinates. As generalized coordinates for nugget chains several different choices are normally possible. In general, the relation between the deterministic generalized velocities, $\dot{q}_s^{(n)}$, and the deterministic generalized fluiddynamic forces $\mathcal{F}_t^{(fd)}$ for a nugget chain is given as $(s = 1, 2, \ldots, d)$

$$\dot{q}_s^{(n)} - \dot{q}_s^{(\kappa)} := -\sum_{t=1}^{d} \mu_{st}^{(n)} \, \mathcal{F}_t^{(fd)}, \tag{10}$$

where d is the number of degrees of freedom and $\dot{q}_s^{(\kappa)}$ expresses the fluid flow field in terms of a generalized velocity. The generalized mobility tensor $(\mu_{st}^{(n)})$ is symmetric and positive semidefinite.

3. Kinetic Energy

The total kinetic energy for both nugget-spring and nugget chains consisting of N rigid nuggets equals [16]

$$\mathcal{K}^{(tot)} = \frac{1}{2} \sum_{\mu=1}^{N} \dot{\vec{r}}_\mu^* \cdot \overset{\Rightarrow}{\boldsymbol{m}}_\mu^* (\vec{r}_\mu^*) \cdot \dot{\vec{r}}_\mu^*, \tag{11}$$

where the spatial mass tensor $\overset{\Rightarrow}{\boldsymbol{m}}_\mu^*$ is block diagonal

$$\overset{\Rightarrow}{\boldsymbol{m}}_\mu^* (\vec{r}_\mu^*) := \begin{pmatrix} m_\mu \overset{\Rightarrow}{\boldsymbol{\delta}} & \overset{\Rightarrow}{\boldsymbol{0}} \\ \overset{\Rightarrow}{\boldsymbol{0}} & \overset{\Rightarrow}{\boldsymbol{m}}_\mu^{(rot)} (\vec{r}_\mu^{(rot)}) \end{pmatrix}. \tag{12}$$

Matrix $\overset{\Rightarrow}{\boldsymbol{\delta}}$ is the 3x3 unit matrix, and the mass of nugget μ is denoted m_μ. The rotational mass tensor is given by [16]

$$\overset{\Rightarrow}{\boldsymbol{m}}_\mu^{(rot)} (\vec{r}_\mu^{(rot)}) := \left(\frac{\partial \vec{\omega}_\mu}{\partial \dot{\vec{r}}_\mu^{(rot)}} \right) \cdot \overset{\Rightarrow}{\boldsymbol{I}}_\mu (\vec{r}_\mu^{(rot)}) \cdot \left(\frac{\partial \vec{\omega}_\mu}{\partial \dot{\vec{r}}_\mu^{(rot)}} \right)^{\mathrm{T}}, \tag{13}$$

where $\vec{\omega}_\mu$ is the Cartesian angular velocity and $\overset{\Rightarrow}{\boldsymbol{I}}_\mu (\vec{r}_\mu^{(rot)})$ is the moment of inertia tensor of nugget μ. For nugget-spring chains the components of the grand spatial vector $\vec{r}^*$ constitute a set of generalized coordinates. This is not the case for nugget chains, and the kinetic energy for nugget chains expressed in terms of the generalized coordinates $\vec{q}^{(n)}$ reads

$$\mathcal{K}^{(\text{tot},n)} = \frac{1}{2} \sum_{s,t=1}^{d} m_{st}^{(n)}(\vec{q}^{(n)}) \; \dot{q}_s^{(n)} \; \dot{q}_t^{(n)}, \tag{14}$$

where the generalized mass tensor is symmetric and given as $(s, t = 1, 2, \ldots, d)$

$$m_{st}^{(n)} := \sum_{\mu=1}^{N} \left(\frac{\partial \vec{r}_\mu^{*(n)}}{\partial q_s^{(n)}} \right) \cdot \overset{\Rightarrow}{m}_\mu^* \cdot \left(\frac{\partial \vec{r}_\mu^{*(n)}}{\partial q_t^{(n)}} \right)^T. \tag{15}$$

The numerical values of the components of the mass tensors depend on both the details of the nugget surface topography and the mass density distribution within each nugget. All mass tensors are positive semidefinite. We shall in the following see that the determinant of the generalized mass tensor for both nugget and nugget-spring chains plays an important role in describing polymer chain dynamics.

Finding the analytic expression for $|\overset{\Rightarrow}{m}^{(n)}|$ for a given nugget chain using Eq. (15) tends to be prohibitively complex even when the final analytic expression is quite simple. An alternative approach to finding the analytic expression for $|\overset{\Rightarrow}{m}^{(n)}|$ is available [16]

$$|\overset{\Rightarrow}{m}^{(n)}| = |\overset{\Rightarrow}{m}^*| \left| \frac{\partial \vec{r}^*}{\partial \vec{q}^{(ns)}} \right|^2 \left| \left(\overset{\Rightarrow}{m}^{(g)} \right)^{-1} \right|, \tag{16}$$

where $\left| \frac{\partial \vec{r}^*}{\partial \vec{q}^{(ns)}} \right|$ is the Jacobian of the vector transformation $\vec{q}^{(ns)} \to \vec{r}^*$, and $(k, l = 1, 2, \ldots, 6N - d)$

$$\left| \left(\overset{\Rightarrow}{m}^{(g)} \right)^{-1} \right| := \left| \sum_{\mu=1}^{N} \left(\frac{\partial g_k}{\partial \vec{r}_\mu^*} \right)^T \cdot \left(\overset{\Rightarrow}{m}_\mu^* \right)^{-1} \cdot \left(\frac{\partial g_l}{\partial \vec{r}_\mu^*} \right) \right|. \tag{17}$$

4. Thermodynamic Equilibrium

We will here assume that each polymer chain is in thermodynamic equilibrium with the fluid in which it is embedded and that the system is isothermal. Using standard methods from statistical physics involving contraction over the conjugated momenta in phase-space it can be shown [10] that the equilibrium conformation-space probability density for an ensemble of identical nugget or nugget-spring chains is independent of the polymer chain mobility tensor and reads

$$
\begin{aligned}
p^{(\mathrm{eq})}(\vec{q}) &= \frac{\exp\left\{-[\Phi^{(\mathrm{tot})}(\vec{q})]/(k_{\mathrm{B}}T)\right\}|\overset{\Rightarrow}{\boldsymbol{m}}^{(\mathrm{q})}|^{1/2}}{\int \exp\left\{-[\Phi^{(\mathrm{tot})}(\vec{q})]/(k_{\mathrm{B}}T)\right\}|\overset{\Rightarrow}{\boldsymbol{m}}^{(\mathrm{q})}|^{1/2}\,\mathrm{d}\vec{q}} \\
&= \frac{\exp\left\{-[\Phi^{(\mathrm{tot})}(\vec{q})+\Phi^{(m)}(\vec{q})]/(k_{\mathrm{B}}T)\right\}}{\int \exp\left\{-[\Phi^{(\mathrm{tot})}(\vec{q})+\Phi^{(m)}(\vec{q})]/(k_{\mathrm{B}}T)\right\}\mathrm{d}\vec{q}},
\end{aligned} \tag{18}
$$

where k_{B} equals the Boltzmann constant, T is absolute temperature, $\vec{q}$ is any choice of generalized coordinates, $\Phi^{(\mathrm{tot})}(\vec{q})$ is the total potential energy and

$$
\Phi^{(m)}(\vec{q}) := -\frac{1}{2}k_{\mathrm{B}}T \, \ln|\overset{\Rightarrow}{\boldsymbol{m}}^{(\mathrm{q})}|. \tag{19}
$$

The formal force derived from the latter equivalent potential reads ($s = 1, 2, \ldots, d$)

$$
\mathcal{F}_s^{(m)}(\vec{q}) := -\frac{\partial}{\partial q_s}\Phi^{(m)}(\vec{q}). \tag{20}
$$

This force is commonly referred to as the metric force.

Use of the results given in Eqs. (16) and (20) yields that the metric force of nugget chains can be split into the following two components

$$
\mathcal{F}_s^{(m)}(\vec{q}) := \mathcal{F}_s^{(m,\mathrm{rot})}(\vec{q}) + \mathcal{F}_s^{(m,g)}(\vec{q}), \tag{21}
$$

$$
\mathcal{F}_s^{(m,\mathrm{rot})}(\vec{q}) := \frac{1}{2}k_{\mathrm{B}}T\frac{\partial}{\partial q_s}\ln|\overset{\Rightarrow}{\boldsymbol{m}}^{*}|, \tag{22}
$$

$$
\mathcal{F}_s^{(m,g)}(\vec{q}) := k_{\mathrm{B}}T\frac{\partial}{\partial q_s}\ln\left|\frac{\partial\vec{r}^{*}}{\partial\vec{q}^{(\mathrm{ns})}}\right| + \frac{1}{2}k_{\mathrm{B}}T\frac{\partial}{\partial q_s}\ln|(\overset{\Rightarrow}{\boldsymbol{m}}^{(g)})^{-1}|. \tag{23}
$$

For the special case when the generalized coordinates of the nugget chain constitute a subset of the components of the grand spatial vector, i.e. $\vec{r}^{*} := \{\vec{q}^{(\mathrm{n})}, \vec{r}_2^{(\mathrm{trans})}, \vec{r}_3^{(\mathrm{trans})}, \ldots, \vec{r}_N^{(\mathrm{trans})}\}$, the Jacobian $|\frac{\partial\vec{r}^{*}}{\partial\vec{q}^{(\mathrm{ns})}}| = 1$, and the first term in the expression for $\mathcal{F}_s^{(m,g)}(\vec{q})$ vanishes.

The component $\mathcal{F}_s^{(m,\mathrm{rot})}$ of the metric force appears because of nugget orientation while the second part $\mathcal{F}_s^{(m,g)}$ is due to the presence of holonomic constraints (ball-socket joints) and the associated reduction in the number of degrees of freedom. The metric force $\mathcal{F}_s^{(m,\mathrm{rot})}$ is independent of the mass distribution within the nuggets. This is in general not the case for $\mathcal{F}_s^{(m,g)}$.

5. Kinetic Theory

After contraction of the phase-space conservation equation (the Liouville equation) over the conjugated momenta of biopolymers embedded in a fluid the following generalized conformation-space conservation equation appears [10]

$$\frac{\partial}{\partial t} p(\vec{q}, t) = -\frac{\partial}{\partial \vec{q}} \cdot \left\{ [\![\dot{\vec{q}}]\!] \ p(\vec{q}, t) \right\}, \tag{24}$$

where $p(\vec{q}, t)$ is the generalized conformation-space probability density and $[\![\dot{\vec{q}}]\!]$ denotes the momentum averaged generalized velocity. We will in the following only consider time scales longer than the momentum relaxation times for the fluid, i.e. we limit our analysis to the diffusion time domain. We will further assume that the relations between $[\![\dot{\vec{q}}]\!]$ and the associated forces for nugget-spring and nugget chains are given by Eqs. (3) and (10), respectively.

Equation (4) gives the force balance when the non-deterministic forces and inertia can be ignored. The effects on polymer chain dynamics due to the forces associated with thermally induced fluiddynamic fluctuations can for any choice of generalized coordinates be taken into account by adding the time averaged Brownian force [10]

$$\mathcal{F}_t^{(B)} := -k_B T \frac{\partial}{\partial q_t} \left\{ \ln \left[\frac{p(\vec{q}, t)}{\sqrt{|\overset{\Rightarrow}{\boldsymbol{m}}{}^{(q)}|}} \right] \right\}. \tag{25}$$

This ensures that the equilibrium probability density given in Eq. (18) is a stationary solution of Eq. (24) for all choices of $\overset{\Rightarrow}{\boldsymbol{\mu}}{}^*$ and $(\mu_{st}^{(n)})$ which are positive semidefinite. Because we here are concerned only with polymer dynamics in the diffusion time domain it suffices to make use of the steady state fluiddynamic forces [17].

Use of Eqs. (4), (10), (20), (24) and (25) yields the following generalized conformation-space diffusion (Fokker-Planck) equation for nugget chains

$$\frac{\partial}{\partial t} p(\vec{q}^{(n)}, t) = -\sum_{s=1}^{d} \frac{\partial}{\partial q_s^{(n)}} \left\{ \sum_{t=1}^{d} \mu_{st}^{(n)} \left[\mathcal{F}_t^{(\kappa)} + \mathcal{F}_t^{(\Phi)} + \mathcal{F}_t^{(e)} + \mathcal{F}_t^{(m,rot)} \right. \right.$$

$$\left. + \mathcal{F}_t^{(m,g)} \right] p(\vec{q}^{(n)}, t) + k_B T \left[\sum_{t=1}^{d} \frac{\partial}{\partial q_t^{(n)}} \mu_{st}^{(n)} \right] p(\vec{q}^{(n)}, t) \right\}$$

$$+ k_B T \sum_{s,t=1}^{d} \frac{\partial}{\partial q_s^{(n)}} \frac{\partial}{\partial q_t^{(n)}} \left\{ \mu_{st}^{(n)} \ p(\vec{q}^{(n)}, t) \right\}. \tag{26}$$

Changing $(\mu_{st}^{(n)})$ into $\overset{\Rightarrow*(\mathrm{ns})}{\mu}$, $\vec{q}^{(n)}$ into $\vec{r}^*$ and $\vec{\mathcal{F}}$ into $\vec{F}^*$ in the latter parabolic partial differential equation yields the conformation-space diffusion equation for nugget-spring chains.

Except for the simplest experimental conditions and the most idealized polymer models, the conformation-space diffusion equation can not be solved analytically. An additional difficulty arises because parabolic partial differential equations can in general not be solved readily using numerical methods when the number of degrees of freedom is large. For most polymer systems the only feasible approach is therefore to instead solve the equivalent stochastic differential equation employing numerical Brownian dynamics simulations. This is the topic of the next section.

6. Stochastic Differential Equations

A stochastic differential equation and a Fokker-Planck equation are classified as weakly equivalent when they describe Markov processes with the same drift and diffusion [11]. For nugget-spring chains with finite $H^{(s)}$ the following Itô stochastic differential equation is weakly equivalent to the diffusion equation for such chains [18]

$$
d\vec{r}_\mu^{*(\mathrm{ns})} := \vec{v}^{*(\kappa)}(\vec{r}_\mu^*)\, dt + \sum_{\nu=1}^{N} \overset{\Rightarrow*(\mathrm{ns})}{\mu}_{\mu\nu} \cdot \left[\vec{F}_\nu^{*(\Phi)} + \vec{F}_\nu^{*(\mathrm{e})} + \vec{F}_\nu^{*(\mathrm{m,rot})} \right] dt
$$

$$
+ k_B T \sum_{\nu=1}^{N} \frac{\partial}{\partial \vec{r}_\nu^*} \cdot \overset{\Rightarrow*(\mathrm{ns})}{\mu}_{\mu\nu}\, dt + \sqrt{2k_B T} \sum_{\nu=1}^{N} \overset{\Rightarrow*(\mathrm{ns})}{B}_{\mu\nu} \cdot d\vec{W}_\nu^*, \qquad (27)
$$

where $\sum_{\eta=1}^{N} \overset{\Rightarrow*(\mathrm{ns})}{B}_{\mu\eta} \cdot \overset{\Rightarrow*(\mathrm{ns})\mathrm{T}}{B}_{\nu\eta} := \overset{\Rightarrow*(\mathrm{ns})}{\mu}_{\mu\nu}$, and $d\vec{W}_\nu^*$ is a 6-dimensional Wiener process. It is here of interest to note that the stochastic part of the dynamics is expressed in terms of the mobility tensor for deterministic motion. This is in accordance with the fluctuation-dissipation theorem [17].

In the limit $H^{(s)} \to \infty$ the Itô stochastic differential equation associated with the diffusion equation of nugget-spring chains reads [16]

$$
d\vec{r}_\mu^{*(\mathrm{ns},\infty)} := \vec{v}^{*(\kappa)}(\vec{r}_\mu^*)\, dt
$$

$$
+ \sum_{\nu=1}^{N} \left(\sum_{\eta=1}^{N} \overset{\Rightarrow*}{P}_{\mu\eta} \cdot \overset{\Rightarrow*(\mathrm{ns})}{\mu}_{\eta\nu} \right) \cdot \left[\vec{F}_\nu^{*(\Phi)} + \vec{F}_\nu^{*(\mathrm{e})} + \vec{F}_\nu^{*(\mathrm{m,rot})} \right] dt
$$

$$
+ k_B T \sum_{\nu=1}^{N} \frac{\partial}{\partial \vec{r}_\nu^*} \cdot \left(\sum_{\eta=1}^{N} \overset{\Rightarrow*}{P}_{\mu\eta} \cdot \overset{\Rightarrow*(\mathrm{ns})}{\mu}_{\eta\nu} \right) dt
$$

$$
+ \sqrt{2k_B T} \sum_{\eta,\nu=1}^{N} \overset{\Rightarrow*}{P}_{\mu\eta} \cdot \overset{\Rightarrow*(\mathrm{ns})}{B}_{\eta\nu} \cdot d\vec{W}_\nu^*, \qquad (28)
$$

because $\vec{r}^*$ constitutes a proper choice of generalized coordinates also in the limit $H^{(s)} \to \infty$.

For nugget chains the Itô stochastic differential equation corresponding to the conformation-space diffusion Eq. (26) reads

$$
dq_s^{(n)} = \sum_{t=1}^{d} \mu_{st}^{(n)} [\mathcal{F}_t^{(\kappa)} + \mathcal{F}_t^{(\Phi)} + \mathcal{F}_t^{(e)} + \mathcal{F}_t^{(m,rot)} + \mathcal{F}_t^{(m,g)}] \, dt
$$

$$
+ k_B T \sum_{t=1}^{d} \frac{\partial}{\partial q_t^{(n)}} \mu_{st}^{(n)} \, dt + \sqrt{2k_B T} \sum_{t=1}^{d} B_{st}^{(n)} \, dW_t, \tag{29}
$$

where

$$
\sum_{u=1}^{d} B_{su}^{(n)} B_{ut}^{(n)} = \mu_{st}^{(n)}. \tag{30}
$$

The latter two equations are valid for any choice of generalized coordinates, but it is often helpful to choose generalized coordinates that constitute a subset of the components of the spatial coordinates.

In Eq. (29) the Itô stochastic differential equation for nugget chains is given in terms of generalized coordinates. This equation can be transformed into a Itô stochastic differential equation expressed only in terms of spatial coordinates by first employing

$$
d\vec{r}_\mu^{*(n)} = \sum_{s=1}^{d} \frac{\partial \vec{r}_\mu^{*(n)}}{\partial q_s^{(n)}} \, dq_s^{(n)} + \frac{1}{2} \sum_{s,t=1}^{d} \frac{\partial^2 \vec{r}_\mu^{*(n)}}{\partial q_s^{(n)} \partial q_t^{(n)}} \, dq_s^{(n)} dq_t^{(n)} + ... \tag{31}
$$

and only include terms proportional to $\Delta t^{1/2}$ and Δt [11]. For nugget chains this yields that [16]

$$
d\vec{r}_\mu^{*(n)} = \sum_{\nu=1}^{N} \left(\sum_{\eta=1}^{N} \overset{\Rightarrow}{\vec{P}}_{\mu\eta}^{*} \cdot \overset{\Rightarrow}{\vec{\mu}}_{\eta\nu}^{*(ns)} \right) \cdot \left[\vec{F}_\nu^{*(\kappa)} + \vec{F}_\nu^{*(\Phi)} + \vec{F}_\nu^{*(e)} \right.
$$

$$
\left. + \vec{F}_\nu^{*(m,rot)} + \vec{F}_\nu^{*(m,g)} \right] dt + k_B T \sum_{\nu=1}^{N} \frac{\partial}{\partial \vec{r}_\nu^*} \cdot \left(\sum_{\eta=1}^{N} \overset{\Rightarrow}{\vec{P}}_{\mu\eta}^{*} \cdot \overset{\Rightarrow}{\vec{\mu}}_{\eta\nu}^{*(ns)} \right) dt
$$

$$
+ \sqrt{2k_B T} \sum_{\eta,\nu=1}^{N} \overset{\Rightarrow}{\vec{P}}_{\mu\eta}^{*} \cdot \overset{\Rightarrow}{\vec{B}}_{\eta\nu}^{*(ns)} \cdot d\vec{W}_\nu^*, \tag{32}
$$

where

$$
\vec{F}_\nu^{*(m,g)} = \sum_{t=1}^{d} \frac{\partial q_t^{(n)}}{\partial \vec{r}_\nu^*} \, \mathcal{F}_t^{(m,g)}. \tag{33}
$$

The only difference between Eqs. (28) and (32) is the presence of a non-zero metric force $\vec{F}_\nu^{*(\mathrm{m,g})}$ in the latter equation. This means that in absence of any potentials the only difference between Eqs. (28) and (32) is that the angular orientation of the nuggets in the latter case is correlated. Equations (29) and (32) constitute alternative, but equivalent equations for describing the stochastic dynamics of nugget chains. When $\mathcal{F}_t^{(\mathrm{m,g})} := 0$ in Eq. (29) this equation and Eq. (28) constitute alternative, but equivalent equations for describing the stochastic dynamics of nugget-spring chains in the limit $H^{(\mathrm{s})} \to \infty$.

7. Numerical Algorithms

The Itô stochastic differential equation for nugget-spring chains with finite $H^{(\mathrm{s})}$, Eq. (27), and nugget chains with constraints described using generalized coordinates, Eq. (29), can be integrated readily using the standard Euler-Maruyama integration scheme [11]. For nugget chains the resulting numerical algorithm reads

$$\Delta q_s^{(\mathrm{n})} = \sum_{t=1}^{d} \mu_{st}^{(\mathrm{n})} \left[\mathcal{F}_t^{(\kappa)} + \mathcal{F}_t^{(\Phi)} + \mathcal{F}_t^{(\mathrm{e})} + \mathcal{F}_t^{(\mathrm{m,rot})} + \mathcal{F}_t^{(\mathrm{m,g})} \right] \Delta t$$

$$+ k_{\mathrm{B}}T \sum_{t=1}^{d} \frac{\partial}{\partial q_t^{(\mathrm{n})}} \mu_{st}^{(\mathrm{n})} \, \Delta t + \sqrt{2k_{\mathrm{B}}T} \sum_{t=1}^{d} B_{st}^{(\mathrm{n})} \, \Delta W_t. \tag{34}$$

The matrix $(B_{st}^{(\mathrm{n})})$ is calculated by Cholesky decomposition of $(\mu_{st}^{(\mathrm{n})})$. This algorithm and the corresponding one for nugget-spring chains constitute the natural basis for Brownian dynamics simulations of systems containing these polymer chains.

The Itô stochastic differential equation for nugget-spring chains in the limit $H^{(\mathrm{s})} \to \infty$, Eq. (28), and nugget chains described using spatial coordinates, Eq. (32), may be simulated using the predictor-corrector algorithm [11], which for nugget chains [14] yields

$$\vec{r}_\mu^{*(\mathrm{n})}(t + \Delta t) = \vec{r}_\mu^{*(\mathrm{n},00)} + \left\{ \vec{v}^{*(\kappa)}(\vec{r}_\mu^*) \right.$$

$$\left. + \sum_{\nu=1}^{N} \overset{\Rightarrow}{\boldsymbol{\mu}}_{\mu\nu}^{*(\mathrm{ns})} \cdot [\vec{F}_\nu^{*(\Phi)} + \vec{F}_\nu^{*(\mathrm{e})} + \vec{F}_\nu^{*(\mathrm{m,rot})} + \vec{F}_\nu^{*(\mathrm{m,g})}] \right\}^{(\mathrm{n},00)} \Delta t$$

$$+ \sqrt{2k_{\mathrm{B}}T} \sum_{\nu=1}^{N} \left[\overset{\Rightarrow}{\boldsymbol{B}}_{\mu\nu}^{*(\mathrm{ns})} \right]^{(\mathrm{n},00)} \cdot \Delta \vec{W}_\nu$$

$$- \sum_{j=1}^{d'} \left[\sum_{\nu=1}^{N} \overset{\Rightarrow}{\boldsymbol{\mu}}_{\mu\nu}^{*(\mathrm{ns})} \cdot \frac{\partial g_j}{\partial \vec{r}_\nu^*} \right]^{(\frac{1}{2})} \gamma_j \, \Delta t, \tag{35}$$

where $d' = 6N - d$.

The latter numerical integration scheme satisfies the Itô stochastic differential equation after two iterations in the evaluation of the Lagrange multipliers, γ_j. These parameters can be identified as the components of the Cartesian constraint or spring forces. Superscript $^{(n,00)}$ denotes that the parameters are evaluated at $\vec{r}^{*(n,00)} := \{\vec{r}_1^{*(n,00)}, \vec{r}_2^{*(n,00)}, ..., \vec{r}_N^{*(n,00)}\}$, which describes the nugget chain conformation at time t. Superscript $^{(\frac{1}{2})}$ denotes that the parameters are evaluated at $\vec{r}^* := \vec{r}^{*(n,00)} + \frac{1}{2}\Delta\vec{r}^{*(ns)}$. The spatial position after the non-constrained step reads

$$\vec{r}_\mu^{*(ns)} = \vec{r}_\mu^{*(n,00)} + \Delta\vec{r}_\mu^{*(ns)}, \tag{36}$$

where

$$\Delta\vec{r}_\mu^{*(ns)} := \left[\vec{v}^{*(\kappa)}(\vec{r}_\mu^*) + \sum_{\nu=1}^{N} \overrightarrow{\mu}_{\mu\nu}^{*(ns)} \cdot (\vec{F}_\nu^{*(\Phi)} + \vec{F}_\nu^{*(e)} + \vec{F}_\nu^{*(m,rot)} \right.$$
$$\left. + \vec{F}_\nu^{*(m,g)})\right]^{(n,00)}\Delta t + \sqrt{2k_B T} \sum_{\nu=1}^{N} \left[\overrightarrow{B}_{\mu\nu}^{*(ns)}\right]^{(n,00)} \cdot \Delta\vec{W}_\nu^*. \tag{37}$$

Provided that the quantitative numerical values of the mobility tensor are available all the algorithms presented above can be used to carry out quantitative non-equilibrium Brownian dynamics simulations of the conformational dynamics of nugget-spring and nugget chains.

8. Conclusions

The work presented here provides several alternative approaches for modeling the conformational dynamics and transport properties of nanoparticles consisting of non-spherical subunits interconnected by conservative forces or holonomic constraints. Only detailed comparison of theoretical predictions and precise experimental results can tell which theoretical model is the most appropriate. By analysing experimental data using the presented theory we can in principle determine all molecular parameters that are included in the theoretical description.

The theoretical foundation discussed here can serve as the formal basis for quantitative equilibrium and non-equilibrium Brownian dynamics simulations of a remarkably wide class of nanoparticles in solution, including segmented biopolymers and liquid crystals.

The remaining task is to establish effective algorithm for calculations of the grand mobility tensor for non-spherical nanoparticles. This may prove to be a major challenge, and the work has just barely started.

9. Acknowledgement

This work was in part supported by grant 138394/410 from the Norwegian Research Council.

References

1. Ermak, D.L. and McCammon, J.A. (1978) Brownian dynamics with hydrodynamic interactions, *J Chem Phys* **69**, 1352-1360.
2. Allison, S.A. and McCammon, J.A. (1984) Transport properties of rigid and flexible macromolecules by Brownian dynamics simulation, *Biopolymers* **23**, 167-187.
3. Garcia de la Torre, J. (1994) Hydrodynamics of segmentally flexible macromolecules, *Eur Biophys J* **23**, 307-322.
4. Kramers, H.A. (1944) Het gedrag van macromoleculen in een stroomende vloeistof, *Physica* **11**, 1-19. English translation: (1946) The behavior of macromolecules in inhomogeneous flow, *J Chem Phys* **14**, 415-424.
5. Kirkwood, J.G. (1967) *Macromolecules*, Gordon and Breach, New York.
6. Rouse, P.E. Jr. (1953) A theory of the linear viscoelastic properties of dilute solutions of coiling polymers, *J Chem Phys* **21**, 1272-1280.
7. Zimm, B.H. (1956) Dynamics of polymer molecules in dilute solution: viscoelasticity, flow birefringence and dielectric loss, *J Chem Phys* **24**, 269-278.
8. Zwanzig, R. (1969) Langevin theory of polymer dynamics in dilute solution, *Adv Chem Phys* **15**, 325-331.
9. Fixman, M. (1978) Simulation of polymer dynamics. I. General theory, *J Chem Phys* **69**, 1527-1537.
10. Bird, R.B., Curtiss, C.F., Armstrong, R.C. and Hassager, O. (1987) *Dynamics of Polymeric Liquids, vol. 2, Kinetic Theory*, 2nd edn., Wiley, New York.
11. Öttinger, H.C. (1996) *Stochastic Processes in Polymeric Fluids*, Springer, Berlin.
12. Goldstein, H. (1980) *Classical Mechanics*, 2nd edn., Addison-Wesley, Reading, MA.
13. Naess, S.N., Adland, H.M., Mikkelsen, A. and Elgsaeter, A. (2001) Brownian dynamics simulation of rigid bodies and segmented polymer chains. Use of Cartesian rotation vectors as the generalized coordinates describing angular orientations, *Physica A* **294**, 323-339.
14. Klaveness, E. and Elgsaeter, A. (1999) Brownian dynamics of bead-rod-nugget-spring polymer chains with hydrodynamic interactions, *J Chem Phys* **110**, 11608-11615.
15. Nyland, G.H., Sjetne, P., Mikkelsen, A. and Elgsaeter, A. (1996) Brownian dynamics simulation of needle chains, *J Chem Phys* **105**, 1198-1207.
16. Naess, S.N. and Elgsaeter, A. (2002) Brownian Dynamics of Segmented Biopolymers: A Formal Theory and Numerical Simulations, *Macromol. Theory Simul.* **11**, 913-923.
17. Kubo, R., Toda, M. and Hashitsume, N. (1985) *Statistical Physics II, Nonequilibrium Statistical Mechanics*, Springer, Berlin.
18. Mikkelsen, A., Knudsen, K.D. and Elgsaeter, A. (1998) Brownian dynamics simulation of needle-spring chains, *Physica A* **253**, 66-76.

CYTOKINESIS: THE INITIAL LINEAR PHASE CROSSES OVER TO A MULTIPLICITY OF NON-LINEAR ENDINGS

Biphasic cytokinesis and cooperative single cell reproduction

David Biron
Department of Physics of Complex Systems
Weizmann Institute of Science

Pazit Libros
Department of Biological Chemistry
Weizmann Institute of Science

Dror Sagi
Department of Physics of Complex Systems
Weizmann Institute of Science

David Mirelman
Department of Biological Chemistry
Weizmann Institute of Science

Elisha Moses
Department of Physics of Complex Systems
Weizmann Institute of Science

Abstract   We investigate the final stage of cytokinesis in two types of amoeba, pointing out the existence of *biphasic* furrow contraction. The first phase is characterized by a constant contraction rate, is better studied, and seems universal to a large extent. The second phase is more diverse. In *Dictyostelium discoideum* the transition involves a change in the rate of contraction, and occurs when the width of the cleavage furrow is

A.T. Skjeltorp and A.V. Belushkin (eds.), Forces, Growth and Form in Soft Condensed Matter: At the Interface between Physics and Biology, 217-234.
© 2004 *Kluwer Academic Publishers. Printed in the Netherlands.*

comparable to the height of the cell. In *Entamoeba invadens* the contractile ring carries the cell through the first phase, but cannot complete the second stage of cytokinesis. As a result, a cooperative mechanism has evolved in that organism, where a neighboring amoeba performs directed motion towards the dividing cell, and physically causes separation by means of extending a pseudopod. We expand here on a previous report of this novel chemotactic signaling mechanism.

Keywords: Biphasic cytokinesis, midwife, Dictyostelium discoideum, Entamoeba invadens

1. Introduction

Cytokinesis is the last step in cell division, during which the physical separation of a mitotic cell into two daughter cells is achieved. In the process a constriction appears in the cell circumference, perpendicular to the mitotic spindle and at mid-cell. An acto-myosin contractile ring attached to the inner surface of the cell membrane deepens the cleavage furrow to achieve daughter cell separation. In order to achieve a symmetric partitioning this process must be highly controlled both spatially and temporally.

Cytokinesis has been, and still is, a very active field of research. Recently there have been several developments covering a wide range of phenomenon implicated in the process. The role of microtubules (e.g. the mitotic spindle) is better understood, and it is now known that in many eukaryotes midzone microtubule bundles are important for both spatial positioning of the contractile ring and are continuously required for the progression of the cleavage furrow [Bray and White, 1988, Wheatly and Wang, 1996, Fishkind et al., 1996, Eckley et al., 1997, Zang et al., 1997, Neujahr et al., 1998].

Signaling pathways for spatial and temporal regulating of furrow assembly and ingression are continuously being discovered. Rho-mediated signaling, for instance, is required for initiation of cytokinesis in *Drosophila* [Prokopenko et al., 1999], and for recruitment of actin myosin to the furrow in *Xenopus* embryos [Drechsel et al., 1997].

An understanding of the roles of membrane anchoring, membrane dynamics, and the precise role of myosin and other actin binding proteins is also evolving. For example, the ability of myosin II null *D. discoideum* to perform cytokinesis showed that systems where myosin II is redundant indeed exist [Lozanne and Spudich, 1987, Neujahr et al., 1997b, Uyeda and Yumura, 2000]. Reviews of these developing topics can be found in [Robinson and Spudich, 2000, Field et al., 1999, Wolf et al., 1999, Glotzer, 1997].

Finally, the mechanistic design principles of the actin contractile ring have been of interest [Mabuchi, 1986, Pollard et al., 1990, Satterwhite and Pollard, 1992, Fishkind and Wang, 1993, Pelham and Chang, 2002] although a comprehensive mechanistic model has still not emerged. In summary, despite this impressive progress, much remains unclear about the biochemistry, molecular pathways, regulation and mechanics of cytokinesis.

A genetically tractable model system which has been extensively studied is the cellular slime mould, *Dictyostelium Discoideum* [Weber, 2001, Neujahr et al., 1997b, Neujahr et al., 1997a, Gerisch and Weber, 2000]. This system provides a clear example of two distinct types of cell cycle coupled cytokinesis, namely "cytokinesis-A" and "cytokinesis-B" [Zang et al., 1997, Uyeda et al., 2000, Nagasaki et al., 2001, Nagasaki et al., 2002]. Cytokinesis-A is characterized by an adhesion independent, myosin II driven constriction of the cleavage furrow (e.g. as observed in cells in suspension), while cytokinesys-B, observed mainly in myosin II null *Dictyostelium* mutations, exhibits adhesion dependent, myosin II independent separation furrow constriction. The relative contribution of each of the two mechanisms in wild type adherent cells has not been determined.

Other mutations of *Dictyostelium* cells have been utilized to study different aspects of cytokinesis. The small GTPase RacE, for example, was found to be crucial for furrow progression throughout cytokinesis in suspension [Gerald et al., 1998]. GAPA, a RasGTPase-activating protein encoded by the *gap*A gene, was reported to be specifically involved in the completion of cytokinesis [Adachi et al., 1997]. Similarly, depletion of Dynamin A by gene-targeting techniques obstructs the completion of *Dictyostelium* cytokinesis [Wienke et al., 1999].

The dramatic changes of cell shape during division are indicative of the mechanical forces exerted at the cleavage furrow. There are various measurements of a constant rate contraction phase which lasts until the width of the waist connecting the daughter cells is $\sim 1/10$ of the initial cell diameter [Mabuchi, 1994, Schroeder, 1972]. The steady contraction ensures smooth conditions in the process, but as the connection between daughter cells is about to vanish a singularity is produced, making control non trivial.

At this stage the contractile ring is reported to disintegrate [Schroeder, 1972], and the effect of physical properties of the connecting waist (e.g. elasticity and viscosity) on the contraction becomes increasingly dominant. The contraction therefore enters a second, non linear phase, which has remained particularly enigmatic [Robinson and Spudich, 2000]. While the first part of our paper deals with the biphasic structure of cytokinesis, in the second part we proceed to show what happens in the second phase of the particular case of *E. invadens*.

Entamoeba invadens is a highly motile amoebic parasite originating from reptilian intestines, used as a model for the study of encystation by the human pathogen *Entamoeba histolytica* [Wang et al., 2003]. As we have demonstrated previously [Biron et al., 2001] the *E. invadens* daughter cells can complete cytokinesis by becoming motile, pulling apart from each other and stretching the connecting tether until it breaks. Alternatively, they can cooperate by employing a chemotactic mechanism, i.e. a neighboring cell follows the concentration gradient of a chemical secreted by the dividing cell and physically separates it into two viable daughter cells.

2. Materials and Methods

2.1 Cells

E. invadens was grown at $25 - 27^{\circ}C$ in air tight 42 ml plastic flasks containing TYI-S33 medium (870 ml nanopure water, 20 g trypticase, 10 g yeast extract, 10 g glucose, 2 g NaCl, 1 g K_2HPO_4, 0.6 g KH_2PO_4, 1 g cystein hydrochloride, 0.2 g ascorbic acid, $22.8mg$ ferric ammonium citrate, 130 ml heat inactivated bovine calf serum, 30 ml Diamond vitamin tween 80 solution $40X$ from JRH Biosciences). Observing the cells at $100X$ magnification was enabled by making a hole on the bottom wide face of the flask and sealing it with a thin cover slip glued with paraffin wax.

D. discoideum was grown at $24^{\circ}C$ in petri dishes containing HL5 medium (14.3 g peptone, 7.15 g yeast extract, 18 g maltose, 0.64 g Na_2HPO_4 and 0.49 g KH_2PO_4 per liter, pH 7.0). Using $40X$ magnification required replacing part of the bottom face of the dish with a cover slip, as described above. While motility stops concurrently with division, the observation was initially obscured by the detachment of the connecting section between the two daughter cells from the substrate, and a subsequent loss of focus. Hence, the cells where overlayed with a 2% agarose sheet following the agar-overlay technique [Yumura and Fukui, 1985, Fukui and Inou, 1991]. This allows imaging of the cleavage furrow in the amoeba during the full division time, while allowing the amoeba to live normally between divisions.

2.2 Chemotaxis assay

A $3 - 4$ cm^2 hole was made on the top wide face of the flask (i.e. above the face where the cells were plated). This hole was fitted with a glass cover that was secured in place by paraffin wax prior to plating of cells. Since *E. invadens* are aerotolerant anaerobes, it was essential to preserve their reduced oxygen environment throughout each experiment. Removing the glass cover and then quickly filling the hole with mineral oil shortly before each experiment started enabled us to easily insert glass pipettes into the flask while maintaining a low rate of air diffusion into the medium. A glass pipette, $10-15$ μm in diameter at the tip, was used for locally aspirating and discharging the assayed medium as described in [Biron et al., 2001]. The pipette was held in a fixed position, whereas the flask was placed on a moving stage mounted on a Zeiss Axiovert 135TV inverted microscope. The position of the stage in the horizontal plane was manually controlled by a pair of motorized actuators. All experiments were recorded using a CCD camera at video rate.

2.3 Collection of medium for chemotaxis assays

We collected small volumes ($\sim 10\,pl$) of attractant containing liquid by aspirating medium from the vicinity of the furrow of a dividing amoeba, using a glass micro-pipette with a tip diameter of $5 - 10\mu m$. We also took $1 - 1.5$ ml samples of fluid from the bottom of a $4-$day old flask with an (almost) confluent culture. At this stage the culture is peaking in its division cycle, with over 10^6 divisions per flask per day. This fluid was either directly assayed for attractive activity or subjected to size analysis (e.g. Millipore centricon filtering and SDS-PAGE analysis). The active fraction was further chemically treated as described below.

3. Results

3.1 Cytokinesis in wild type *D. discoideum* is biphasic

The accumulated data from 15 wild type cell divisions in *D. discoideum* is shown in Fig. 1. We show only cases where no motile activity interfered with division, and discuss motile separation below. We plot the behavior of the furrow width (D_m) as a function of $\tau = (t_f - t)$, where t_f is the time of complete separation. A clear transition is observed at the critical time $t_c \sim t_f - 70$ seconds.

The furrow width exhibits two regimes which can be characterized by two different values of the exponent α in the power law $D_m \propto \tau^\alpha$. Separating the two time regimes, fitting them to a power law behavior and then averaging exponents over the 15 divisions gives $\alpha_1 = 1.16 \pm 0.2$ for the initial stage, and $\alpha_2 = 0.65 \pm 0.1$ for the final stage of separation. Previous measurements of the cleavage furrow as a function of

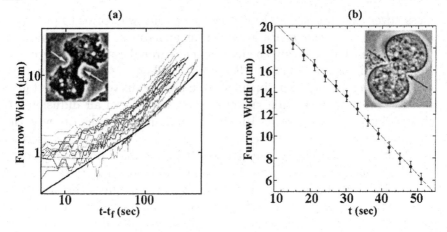

Figure 1a. A log-log plot of *D. discoideum* furrow width, D_m, (under an agar overlay). At time t_f separation of the two daughter cells is completed.

Figure 1b. *E. invadens* furrow width, D_m, as a function of time. At time $t = 0$ furrowing begins.

time have been carried out, for example on dividing eggs of sea urchins [Mabuchi, 1994], and the width of the cleavage furrow was shown to decrease linearly ($\alpha \approx 1$) for D_m was 10% − 20% of the original cell diameter. However, the final stage was not previously presented, e.g. the data presented in [Mabuchi, 1994] does not reach full separation.

A different type of division, in which the two daughter cells regained motility before complete separation in a fashion resembling cytokinesis-B [Robinson and Spudich, 2000, Neujahr et al., 1997b], was recorded for ∼ 20% of the *D. discoideum* divisions under an agarose sheet. At the final stages of these cases, the daughter cells would move in opposite directions, parallel to the connecting tether, stretching it by applying tension forces along its long axis until it ripped. This process was typically slower than cytokinesis-A dominated separation. We conclude that *D. discoideum* furrowing exhibits two regimes: a linear regime and a second which can be either non-motile and accelerating, or motile and prolonged.

3.2 Cytokinesis in *E. invadens*

We have measured the dynamics of cytokinesis of *E. invadens*. There also we observe a linear initial stage, but contrary to the case of *D. discoideum* − we never saw a non-motile second phase in *invadens*. Instead, a transition to motile separation always occurred at the end of the linear stage. Fig. 1 shows the furrow dynamics up to the time when

$D_m(t) \approx 6 \ \mu m \sim 0.25 D_m(0)$. The conclusion of the process could not be measured since *E. invadens* did not enter the same non-linear part of cytokinesis. Instead, the contraction of the ring halted, sometimes for several minutes.

As we reported previously [Biron et al., 2001], we were able to discern three alternative processes that continued beyond the arrest in the cleavage process. In all cases cytokinesis-A was abandoned after the pause. In a majority of the cases cytokinesis was either reversed, or resumed by an application of traction forces by the two daughter cells, actively advancing in opposite directions. This resembles the phenotypes of several mutations known to cause cytokinesis defects in other cells, as we discuss in the next section.

3.3   Cooperative behavior – "Midwife" statistics

We have recorded $n = 106$ *invadens* divisions from four different samples, all at a density of about $3 \cdot 10^4$ cells/cm^2 and without an agar overlay. The results are summarized in Fig. 2a. During the linear phase there is no motile activity in any of the daughter cells. As the second stage is approached, the connective cylinder between the two daughter cells detaches from the substrate and cleavage stalls. At this point both daughter cells resume motility and begin to move away from one another, pulling at the tether that keeps them bound to each other.

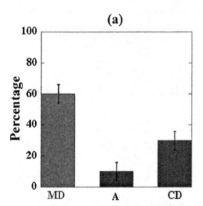 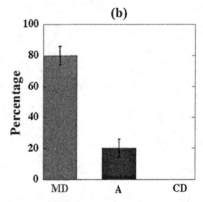

Figure 2a.   The distribution of the three observed terminations of *invadens* cytokinesis. MD – motile divisions; A – abortions; CD – cooperative divisions ("midwives"). The distribution was measured at a cell density of $3 \cdot 10^4$ cells/cm^2.

Figure 2b.   The same distribution measured at a cell density of $5 \cdot 10^3$ cells/cm^2. Midwives were not detected at the lower cell density.

In 63 cases (60 ± 6%) motile separation of the amoeba was observed: the tension on the tether increased as they pulled and it grew longer and narrower until it was cut. Another 11 (10 ± 2%) cases ended in "abortions", where cytokinesis stopped and the amoeba continued to live as a multi-nucleated cell. The cells may then resume division and we have seen division into three or four viable daughter cells. These types of behavior have also been previously observed in the myosin II null mutants of *D. discoideum* [Neujahr et al., 1997b]. A plausible interpretation is that in *E. invadens* the function of the contractile ring is incomplete.

Surprisingly, in the remaining 32 (30±6%) cases a neighboring amoeba intervened in the process of division. This "midwife", as we call it, can travel a long distance (we have measured directed motion of up to 200 μm), usually in a straight trajectory to the dividing amoeba. It aligns itself parallel to the dividing amoeba, sends a pseudopod inbetween the separating amoebae and then severs the connection by moving itself forcibly into the gap and pushing them apart [Biron et al., 2001]. This process is depicted in Fig. 3.

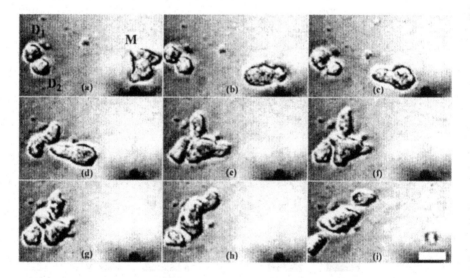

Figure 3. A "midwife" (labeled "M") advancing towards a dividing cell and cutting through the tether connecting the two daughter cells (labeled "D_1" and "D_2") at the final stage of *invadens* cytokinesis. Frames were taken at $t = 0$, 50, 74, 123, 136, 142, 146, 162 and 184 seconds. Scale bar is 10 μm.

3.4   The "midwife" is chemotactic

In order to determine whether chemotaxis accounts for the long directed trajectory that the midwife travels we aspirated $\sim 10\ pl$ of medium from around a dividing amoeba using a micro-pipette (Fig. 4). This medium was then released in a controlled manner near a distant amoeba. In over 50% of the cases a chemotactic response was found, and the amoeba followed the pipette as we moved it away for distances of up to a few hundred micro-meters.

(a)   **(b)**   **(c)**

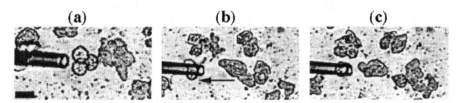

Figure 4a.   Aspirating $\sim 10pl$ medium from the vicinity of a dividing *E. invadens*. Using a micro-pipette, the fluid was then released near far-off amoebae. As cells approached the secreting pipette, it was pulled away from them.

Figure 4b.   An amoeba following the secreting pipette. The arrow shows the direction of movement. This cells followed the retracting pipette for $\sim 200\ \mu m$.

Figure 4c.   The same cell closer to the tip after $20sec$ in which the pipette was held in place. Scale bar is 10 μm.

Movies showing these measurements appear at [Nature, 2001]. In the cases when chemotaxis did not work, we could invariably point out factors of the experiment that could have made it faulty (e.g. the pipette was too thin) but we could not exclude the possibility that only part of the dividing cell population secretes an attractant, or that midwives react only in a specific stage of the cell cycle. Controls such as aspirating near non-dividing amoeba or using regular medium gave no response.

To obtain larger volumes of medium containing attractant, we relied on chemoattractants being long lasting (unless a specific enzyme is targeted to degrade them) [Eisenbach, 2003], and took $1 - 1.5ml$ samples of medium from a flask containing confluent 4-day old culture, as described in Sec. 2. We assayed these samples, and got a strong chemotactic response (Fig. 5). A control experiment using 4-day old medium from the same batch, in which no cells were grown, did not elicit any response.

The assay for chemotaxis was quantified by comparing the velocities of *E. invadens* cells that were suspected to be stimulated by the "midwife" calling signal and non-stimulated cells. Fig. 6 depicts a set of such measurements.

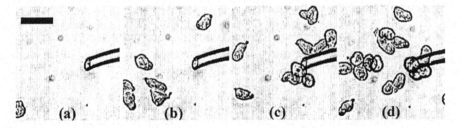

Figure 5. *E. invadens* cells in fresh medium attracted to a stationary micro pipette secreting medium originating from a 4−day old almost confluent culture. Scale bar is 50 μm. $t = 0, 375, 750$ and 1290 *sec*.

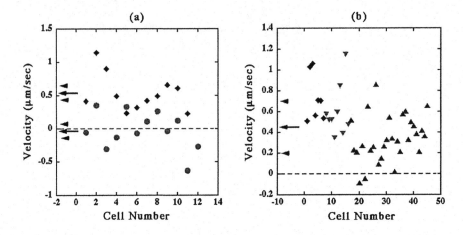

Figure 6a. Diamonds denote velocities of *E. invadens* cells stimulated by a retracting pipette containing medium from the vicinity of a dividing amoeba ($v_{stim} = 0.5 \pm 0.05 \ \mu m/sec$); Circles denote control experiments done in the same fashion but with a pipette filled with plain medium, i.e. cells were not stimulated ($v_{non} = 0 \pm 0.05 \ \mu m/sec$). The arrows show the average velovity $\pm$ one standard deviation.

Figure 6b. Velocities of *E. invadens* cells advancing towards a pipette containing medium extracted from a confluent culture flask. Diamonds denote horizontal velocities towards a retracting pipette. Triangles denote radial velocities towards a stationary pipette. The arrows show the average velovity $\pm$ one standard deviation.

We have found the average stimulated cell velocity to be $v_{stim} = 0.5 \pm 0.05 \ \mu m/sec$, while the non-stimulated cells were randomly diffusing about at $v_{non} = 0 \pm 0.05 \ \mu m/sec$. The significant difference (an order of magnitude above the noise) enabled us to use cell velocities to distinguish directed from random motion. These results compared well to the velocity of directed motion towards a source of high glucose concentration which was $\sim 1~\mu m/sec$ (the exact value depended on various parameters e.g. the glucose concentration, the geometry of the pipette and the initial velocity of the flow out of the pipette).

3.5   Density dependence of cooperativity

There is a strong dependence of the statistical distribution of divisions on factors such as cell density, medium quality etc. In a control experiment we reduced the cell density from $3 \cdot 10^4$ cells/cm^2 to $5 \cdot 10^3$ cells/cm^2 and measured the distribution of cytokinesis terminations (Fig. 2). Such a reduction in cell density increases the average distance between cells ~ 2.5-fold, so that it is $\sim 150\mu m$. As a result, two factors hinder the chemotactic response. When the distance of diffusion exceeds a few hundreds of micro-meters the chemoattractant reaches its target highly diluted. In addition, as we will demonstrate, the time it takes a neighboring cell to respond to the signal becomes too large.

Our preliminary data shows the chemoattractant to have a molecular weight of $50 \pm 10kDa$. A globular molecule of this size will diffuse a distance of $150\mu m$ in an average time of $t_D \approx 2min$ in water at 25^oC. A midwife moving at a velocity of $\sim 0.5\mu m/s$ would travel this distance in $t_T \approx 5min$. Therefore the time it would take for a signal to get to a neighbor cell, and for the cell to respond and approach would be $t_D + t_T \geq \Delta t$, where $\Delta t \sim 5 - 10min$ is the typical time interval between the release of the signal and the completion (in some way) of cytokinesis. This increases the probability that the dividing cell will either complete its separation or abort the process before help arrives.

When the midwife can no longer come to aid separation, the cases which would have concluded in cooperation at higher densities are divided between successful motile separations and abortions. If we consider only the cases in which a "midwife" did not intervene, we can see in Fig. 2, that at the high cell density approximately one in every seven unassisted attempts to divide ends up in an "abortion". At the low cell density, however, as many as one in every five unassisted attempts to divide fails i.e. cooperation increases the ratio of successful motile (independent) separations to abortions.

4.   Discussion

4.1   Cytokinesis in *D. discoideum*

We have measured a transition in the scaling of *D. discoideum* furrow contraction dynamics. After a short transient the furrow contracts (approximately) linearly until its width is of the order of the cell height. It then speeds up ($\alpha = 0.65 \pm 0.1$) towards the end.

Comparing our measurements of wild type cells to the furrow dynamics of mutant $D.$ $discoideum$ suggests that biphasic dynamics, i.e. a linear contraction phase throughout $80 - 90\%$ of the furrow progression, followed by a short non-linear phase is a signature of normal cytokinesis-A in adherent cells. In contrast, data of furrow dynamics of adherent myosin II null cells (i.e. cytokinesis-B), published in [Zang et al., 1997], demonstrates the lack of a linear contraction phase. In fact, the data published does not exhibit a transition between two dynamic regimes, but can be fitted by a single power law, $D_m \propto \tau^\alpha$ with an exponent of $\alpha = 0.55 \pm 0.05$. For comparison, Zang et al. present the furrow dynamics of ΔBLCBS-myosin cells, which are assumed to overexpress myosin. The data shows a linear phase ($\alpha \approx 1$) which is even longer than in typical wild type cells, ranging from a furrow width of $D_m \approx 11\mu m$ to $D_m \approx 1\mu m$.

Another mutation of interest is dynamin A null $D.$ $discoideum$, i. e. $dymA^-$ cells [Wienke et al., 1999]. These cells are able to grow in suspension, form a functional contractile ring during cell division, and pass through all stages of cytokinesis up to the point where only a thin cytoplasmatic bridge is connecting the two daughter cells. The final severing of this bridge is inhibited by lack of dynamin. The published data on $dymA^-$ cells shows a biphasic dynamics, with an initial linear phase and a nonlinear ($\alpha < 1$) second phase.

Finally, the protein GAPA is specifically involved in the completion of cytokinesis [Adachi et al., 1997]. Again, $D.$ $discoideum$ cells lacking GAPA ($gapA^-$) are reported to exhibit normal cytokinesis, including myosin II accumulation in the cleavage furrow, until the step in which the daughter cells should be severed. In light of the cases reported above we can conclude that the presence of a biphasic furrow dynamics qualifies as a quantitative measure of "normal" cytokinesis-A.

4.2   Cytokinesis in $E.$ $invadens$

The linear contraction phase was also measured in $E.$ $invadens$. In this cell, however, the second phase and the conclusion of successful cytokinesis is carried out utilizing motility. In $\sim 30\%$ of the cases the separation mechanism involves the novel phenomenon of cooperation with a neighbor (non dividing) "midwife", recruited by a chemotactic signal.

The arrest in cytokinesis of $E.$ $invadens$ at the end of the linear phase resembles the phenotype of mutations known to obstruct cytokinesis in other cells. It is therefore worthwhile to compare it with other organisms, and specifically with cytokinesis deficient mutants. Following furrow localization and the beginning of constriction, such mutations commonly display an arrest in ingression, and then either reverse cytokinesis to form a multinucleated cell or abandon cytokinesis-A and use motility (in opposing directions) to complete the separation between the daughter cells.

For example, adherent $gapA^-$ cells (on a glass substrate) initiated cytokinesis similarly to wild type cells until the daughter cells were connected by a thin cytoplasmatic bridge [Adachi et al., 1997]. At this stage the bridge remained intact for a long time, and in $\sim$ 50% of the cases cytokinesis was finally reversed. The other cases completed separation using traction forces, reminiscent of what happens in $E.\ invadens$.

Another similar phenotype consists of $D.\ discoideum$ cells lacking racE. These mutant cells could initiate furrowing but then failed to divide in suspension [Gerald et al., 1998]. A third example is the cells of $Caenorhabditis\ elegans$ embryos deprived of ZEN-4 (a homologue of the MKLP1 subfamily of kinesin motor proteins capable of bundling antiparallel microtubules), which were shown to initiate the cytokinetic furrow at the normal time and place. However, their furrow propagation halts prematurely, and after a short pause it retracts, ultimately producing a single multinucleated cell.

In contrast, when a checkpoint control is affected by a mutation in budding yeast the phenotype is different. Act5 is a yeast actin related protein which is necessary for dynein function, and hence is important for spindle orientation [Muhua et al., 1994, Muhua et al., 1998]. Mutant cells often misalign the mitotic spindle, slowing its entrance to the neck between mother cell and budding daughter cell. This delays cytokinesis at a checkpoint control. After the delay, if the spindle enters the neck, cytokinesis is completed normally.

Following our report of the $E.\ invadens$ "midwife", similar cooperative behavior has been observed in other cells. Mechanical intervention in cytokinesis in $D.\ discoideum$ was recently reported and is currently studied [Insall et al., 2001, Nagasaki et al., 2002]. In addition, fibroblast cells on a smooth surface were observed cooperating in a similar fashion [Bornens and Théry, 2002]. To our knowledge, there have been no reported cases of similar "midwiving" behavior in a mammalian extracellular matrix. It would be extremely interesting if this was not restricted to highly motile cells, but could also occur in neighboring cells in tissue.

Acknowledgements

This investigation was supported in part by the Binational Science Foundation under grant no. 2000298, by the Clore Center for Biological Physics, by the EU Research Training Network PHYNECS and by a grant from the Center for Emerging Diseases, Jerusalem. We thank Shoshana Ravid for providing us with the *D. discoideum* and for advice on their culture and observation.

References

[Adachi et al., 1997] Adachi, H., Takahashi, Y., Hasebe, T., Shirouzu, M., Yokoyama, S., and Sutoh, K. (1997). *Dictyostelium* iqgap-related protein specifically involved in the completion of cytokinesis. *J. Cell Bio.*, 137(4):891–898.

[Biron et al., 2001] Biron, D., Libros, P., Sagi, D., Mirelman, D., and Moses, E. (2001). 'midwifes' assist dividing amoebae. *Nature*, 410:430.

[Bornens and Théry, 2002] Bornens, M. and Théry, M. (2002). Private communications.

[Bray and White, 1988] Bray, D. and White, J. G. (1988). Cortical flow in animal cells. *Science*, 239:883–888.

[Drechsel et al., 1997] Drechsel, D. N., Hyman, A. A., Hall, A., and Glotzer, M. (1997). A requirement for rho and cdc42 during cytokinesis in *xenopus* embryos. *Curr. Biol.*, 7(1):12–23.

[Eckley et al., 1997] Eckley, D. M., Ainsztein, A. M., Mackay, A. M., Goldberg, I. G., and Earnshaw, W. C. (1997). Chromosomal proteins and cytokinesis: patterns of cleavage furrow formation and inner centromere protein positioning in mitotic heterokaryons and mid-anaphase cells. *J. Cell Biol.*, 136(6):1169–1183.

[Eisenbach, 2003] Eisenbach, M. (2003). *Chemotaxis*. Imperial College Press, Covent Garden, UK, 1st edition.

[Field et al., 1999] Field, C., Li, R., and Oegema, K. (1999). Cytokinesis in eukaryotes: a mechanistic comparison. *Curr. Opin. Cell Biol.*, 11:68–80.

[Fishkind et al., 1996] Fishkind, D. J., Silverman, J. D., and Wang, Y. L. (1996). Function of spindle microtubules in directing cortical movement and actin filament organization in dividing cultured cells. *J. Cell Sci.*, 109:2041–2051.

[Fishkind and Wang, 1993] Fishkind, D. J. and Wang, Y. (1993). Orientation and three-dimensional organization of actin filaments in dividing cultured cells. *J. Cell Bio.*, 123:837–848.

[Fukui and Inou, 1991] Fukui, Y. and Inou, S. (1991). Cell division in dictyostelium with special emphasis on actomyosin organization in cytokinesis. *Cell Motil. Cytoskel.*, 18:41–54.

[Gerald et al., 1998] Gerald, N., Dai, J., Ting-Beall, H. P., and De-Lozanne, A. (1998). A role for *dictyostelium* race in cortical tension and cleavage furrow progression. *J. Cell Bio.*, 141(2):483–492.

[Gerisch and Weber, 2000] Gerisch, G. and Weber, I. (2000). Cytokinesis without myosin ii. *Curr. Opin. Cell Biol.*, 12:126–132.

[Glotzer, 1997] Glotzer, M. (1997). The mechanism and control of cytokinesis. *Curr. Opin. Cell Biol.*, 9:815–823.

[Insall et al., 2001] Insall, R., Muller-Taubenberger, A., Machesky, L., Kohler, J., Simmeth, E., Atkinson, S. J., Weber, I., and Gerisch, G. (2001). Dynamics of the dictyostelium arp2/3 complex in endocytosis, cytokinesis, and chemotaxis. *Cell Motil. Cytoskeleton*, 50(3):115–128.

[Lozanne and Spudich, 1987] Lozanne, A. De and Spudich, J. A. (1987). Disruption of the *dictyostelium* myosin heavy chain gene by homologous recombination. *Science*, 236:1086–1091.

[Mabuchi, 1986] Mabuchi, I. (1986). Biochemical aspects of cytokinesis. *Int. Rev. cytol.*, 101:175–213.

[Mabuchi, 1994] Mabuchi, I. (1994). Cleavage furrow: timing of emergence of contractile ring actin filaments and establishment of the contractile ring by filament bundling in sea urchin eggs. *J. Cell Sci.*, 107:1853–1862.

[Muhua et al., 1998] Muhua, L., Adames, N. R., Murphy, M. D., Shields, C. R., and Cooper, J. A. (1998). A cytokinetic checkpoint requiring the yeast homologue of an apc-binding protein. *Nature*, 393:487–491.

[Muhua et al., 1994] Muhua, L., Karpova, T. S., and Cooper, J. A. (1994). A yeast actin-related protein homologous to that in vertebrate dynactin complex is important for spindle orientation and nuclear migration. *Cell*, 78:669–679.

[Nagasaki et al., 2002] Nagasaki, A., de Hostos, E. L., and Uyeda, T. Q. P. (2002). Genetic and morphological evidence for two parallel pathways of cell-cycle-coupled cytokinesis in dictyostelium. *Cell Motil. Cytoskel.*, 50(3):2241–2251.

[Nagasaki et al., 2001] Nagasaki, A., Hibi, M., Asano, Y., and Uyeda, T. Q. P. (2001). Gentic approaches to dissect the mechamisms of two distinct pathways of cell cycly-coupled cytokinesis in *dictyostelium*. *Cell Struc. Fun. ?*, 26:585–591.

[Nature, 2001] Nature, Web-Site (2001). Supplementary information v410 430. $http : //www.nature.com/cgi - taf/DynaPage.taf?file = /nature/journal/v410/n6827/abs/410430a0_fs.html$.

[Neujahr et al., 1998] Neujahr, R., Albrecht, R., Kohler, J., Matzner, M., Schwartz, J. M., Westphal, M., and Gerisch, G. (1998). Microtubule-mediated centrosome motility and the positioning of cleavage furrows in multinucleate myosin ii-null cells. *J. Cell Sci.*, 111(9):1227–1240.

[Neujahr et al., 1997a] Neujahr, R., Heizer, C., Albrecht, R., , Ecke, M., Schwartz, J. M., Weber, I., and Gerisch, G. (1997a). Three-dimensional patterns in redistribution of myosin ii and actin in mitotic *dictyostelium* cells. *J. Cell Biol.*, 139(9):1793–1804.

[Neujahr et al., 1997b] Neujahr, R., Heizer, C., and Gerisch, G. (1997b). Myosin ii-independent processes in mitotic cells of dictyostelium discoideum: redistribution of the nuclei, re-arrangement of the actin system and formation of the cleavage furrow. *J. Cell Sci.*, 110(2):123–137.

[Pelham and Chang, 2002] Pelham, R. J. Jr and Chang, F. (2002). Actin dynamics in the contractile ring during cytokinesis in fission yeast. *Nature*, 419:82–86.

[Pollard et al., 1990] Pollard, T. D., Satterwhite, L., , Cisek, L., Corden, J., Sato, M., and Maupin, P. (1990). Actin and myosin biochemistry in relation to cytokinesis. *Ann. NY Acad. Sci.*, 582:120–130.

[Prokopenko et al., 1999] Prokopenko, S. N., Brumby, A., O'Keefy, L., Prior, L., He, Y., Saint, R., and Bellen, H. J. (1999). A putative exchange factor for rho1 gtpase is required for initiation of ctokinesis in *drosophila*. *Genes Dev.*, 13(17):2301–2314.

[Robinson and Spudich, 2000] Robinson, D. N. and Spudich, J. A. (2000). Towards molecular understanding of cytokinesis. *Trends Cell Biol.*, 10:228–237.

[Satterwhite and Pollard, 1992] Satterwhite, L. L. and Pollard, T. D. (1992). Cytokinesis. *Cur. Op. Cell Bio.*, 4:43–52.

[Schroeder, 1972] Schroeder, T. E. (1972). The contractile ring. ii. determining its brief existence, volumetric changes, and vital role in cleaving arbacia eggs. *J. Cell. Bio.*, 53:419–434.

[Uyeda et al., 2000] Uyeda, T. Q. P., Kitayama, C., and Yumura, S. (2000). Myosin ii-independent cytokinesis in *dictyostelium*: its mechanism and implications. *Cell Struc. Func. ?*, 25:1–10.

[Uyeda and Yumura, 2000] Uyeda, T. Q. P. and Yumura, S. (2000). Molecular biological approaches to study myosin functions in cytokinesis of dictyostelium. *Micro. Res. Tech. ?*, 49:136–144.

[Wang et al., 2003] Wang, Z., Samuelson, J., Clarck, C. G., Eichinger, D., Paul, J., Van-Dellen, K., Hall, N., Anderson, I., and Loftus, B. (2003). Gene discovery in the *entamoeba invadens* genome. *Mol. Bio. Para.*, 129(1):23–31.

[Weber, 2001] Weber, I. (2001). On the mechanism of cleavage furrow ingression in *dictyostelium*. *Cell Struc Func.*, 26:577–584.

[Wheatly and Wang, 1996] Wheatly, S. P. and Wang, Y. (1996). Midzone microtubule bundles are continuously required for cytokinesis in cultured epithelial cells. *J. Cell Biol.*, 135:981–989.

[Wienke et al., 1999] Wienke, D. C., Knetsch, M. L. W., Neuhaus, E. M., Reedy, M. C., and Manstein, D. J. (1999). Disruption of a dynamin homologue affects endocytosis, organelle morphology, and cytokinesis in *dictyostelium discoideum*. *Mol. Biol. Cell ?*, 10:225–243.

[Wolf et al., 1999] Wolf, W. A., Chew, T. L., and Chisholm, R. L. (1999). Regulation of cytokinesis. *Cell. Mol. Life Sci.*, 55:108–120.

234

[Yumura and Fukui, 1985] Yumura, S. and Fukui, Y. (1985). Reversible cyclic amp-dependent changes in distribution of myosin thick filaments in dictyostelium. *Nature*, 314:194–196.

[Zang et al., 1997] Zang, J., Cavet, G., Sabry, J. H., Wagner, P., Moores, S. L., and Spudich, J. A. (1997). On the role of myosin-ii in cytokinesis: Division of *dictyostelium* cells under adhesive and nonadhesive conditions. *Mol. Biol. Cell*, 8:2617–2629.

INFORMATION DYNAMICS IN LIVING SYSTEMS

JARLE BREIVIK
Section for Immunotherapy
Norwegian Radium Hospital
University of Oslo
0310 Oslo
Norway

Abstract.
The concept of information unifies physics and biology at a fundamental level. This convergence lies at the bart of current scientific and technological developments, and represents a new and interdisciplinary perspective to several fields of research. With reference to our own work on self-replication and cancer development, this paper aims to demonstrate bow the physical understanding of information sheds new light on the behavior of living systems

1. Introduction

Regardless of which perspective we chose, life always concerns transmission of information genetic information, historical information, verbal information, written information, or digital information. The concepts of life and information are inseparably united, and a scientific understanding of living systems also requires an understanding of the term *information*. So what is information? That is not a trivial question to answer, but alongside tremendous advances in information technology there have also been very important theoretical advances, and the key paradigm lies in the realization that information is physical (1, 2).

Information always has a physical substance, be it words on parer, sound waves in air, nerve impulses in the brain, or electromagnetic signals in space. Information cannot exist in vacuum (2, 3). In popular terms, a structure carries information if it has a pattern that corresponds to that ofanother structure. A key carries information about the lock that it fits, the information on a CD is small indentations corresponding for instance to the note sheets and the sound waves of an opera, and this text is information because it fits presumable pattems in the brain of the author – and hopefully also the reader. In scientific terms, information may be defined as shared entropy structures (4).

This physical understanding of information reflects arecent convergence between thermodynamics and information theory (4). Intriguingly however, modem life science is based ona strikingly similar paradigm. The discovery of nucleotide sequences as the four letter

A.T. Skjeltorp and A.V. Belushkin (eds.), Forces, Growth and Form in Soft Condensed Matter: At
the Interface between Physics and Biology, 235-242.
© 2004 *Kluwer Academic Publishers. Printed in the Netherlands.*

language ofinheritance demonstrates the duality of structure and information in a very explicit manner (5). Genetic information is indeed physical, and unifies physics, informatics, and biology at afundamentalievel.

Even more intriguing, there is also a direct link between this convergence of scientific disciplines and Charles Darwin's theory of evolution. Despite profound controversies, Darwin basically proposed a model for the dynamic interplay between heritable traits and the surrounding environment (6). Today we know that heritable traits are generally conveyed as nucleotide sequences, and Darwinian evolution has become a fundamental principle concerning the dynamicsof genetic information. Genetic information is propagated though space and time by DNA replication, and nucleotide sequences that in a given environment successfully catalyze their ownsynthesis will multiply at the cost of less favorable sequences (7). Self-replicating structures thereby produce informational patterns that are increasingly betler adapted to their environment, and evolution by means of natural selection may be understood in a strictly physical perspective (8).

2. Plastic replicators

This converging understanding of physics, informatics, and biology is in many aspects a new wayof thinking, and therefore both challenging and controversial. In principle, however, it all comes together in the basic dynamics of template-replicating polymers (5, 9). Such structures process information by a simple thermodynamic mechanism, and form the basis for practically all aspects of life science. In order to demonstrate this principle in a simple and intuitive manner, we therefore set out to develop a system of plastic building blocks that spontaneously self-organize into just that template-replicating polymers (10): Based on Watson and Crick's well-defined model of the DNA molecule (5) we de due ed thatthe eoneept of template-replieation could be expressed in terms of monomerie objeets that interaetby two interrelated binding mechanisms in responds to environmental fluetuations in energy (Fig. 1):

	General concept:	Biologic correlate:	Presented system:
a	At least two types of objects (A, B) that form two types of bindings (I and II).	The 4 nucleotides of DNA (A, G, T, C).	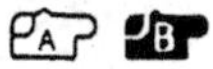
b	Binding I forms specific pairs (A:B) in response to a cyclic variable.	Watson-Crick base pairs (A:T, G:C) in response to the cell cycle.	 in response to temperature cycles.
c	Binding II forms continuous polymers (-A-B-B-A-B-).	Phosphodiester bridges of the DNA backbone.	
d	Binding I more probable than Binding II.	Low vs. high activation energy.	Exposed vs. concealed surfaces.
e	Binding II more stable than Binding I.	Covalent vs. hydrogen bindings.	High T_c vs. low T_c.
f	IF [A₁-B₁ AND A₁:B₂ AND B₂:A₂] THEN A₂-B₂	Protein dependent DNA replication.	

Figure 1. The general concept of template-replicating aperiodie polymers in relation to DNA and the presented ferromagnetic system. Representation of information as strings demands at least two types of monomeric objects (a), represented by the four nucleotides of DNA, and the A and B construets of the model system. Template replication I based on two types of bindings occurring between the objects; Binding I forming complementary pairs in response to acyclic variable (b), and Binding II forming continuous polymers (c). Binding I is more probable than Binding II (d), whereas Binding II is more stable than Binding I (e), corresponding to the relationship between the covalent and the hydrogen bindings of DNA, and the two different magnetic bindings of the model system (Te: Curie temperature). The key to formation and transfer of genetic information lies in the steric and energetic relationship between the two bindings, and may be expressed in terms of conditional logic (t). In simple, monomers that bind to a polymer are joined together in a complementary polymer. This mechanism is dependent on protein catalysis in current biological systems, but may be achieved by simple dynamics as demonstrated by the hinging movement between four connected objects of the model system (*).

Random interaction was achieved by floating the building blocks in a turbulent heat bath, and reversible chemical bindings were modeled by ferromagnetic forces susceptible to temperature fluctuation in the water (10). Self-organization of template-replicating polymers could then be demonstrated (Fig. 2):

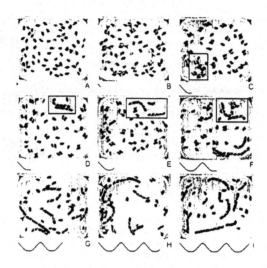

Figure 2. Self-organization of template-replicating polymers through three consecutive thermocycles. Passage through the thermocycles is represented by the extending curve. Key events are indicated. At 60°C all monomeric objects moved independently (A). On lowering the temperature, numerous pair of A (red) and B (blue) objects combined by Binding I (B). A complex comprising a combination of several bindings (I and II) emerged at 15°C (C). The complex flipped into a double-stranded configuration (D). The two strands dissociated after an increase in temperature (E). Dynamic interactions between polymers, monomers and the environment gave rise to secondary and tertiary configurations (F). Long polymers were instable at high temperatures, and breaking of Binding II established an open-ended dynamic (G). A sixmer directed the elongation of a diller (H), resulting a new generation of polymers (I).

This model mechanistically demonstrates self-organization of template-replicating polymers, and simplistically proves that such dynamics are thermodynamically viable. The initial formation of random sequences directs the synthesis of complementary sequences, which logically are not random, and there is an intuitive reduction of uncertainty within the system. Such reduction of uncertainty may be expressed in terms of Shannon's theory of communication (11, 12), and relate directly to the definition of information. Reiterated from above, information is de fine d as shared entropy between two independent structures (2, 4), and that is exactly the result when complementary polymers split apart. Combined, it may thus be stated that this system demonstrates self-organization of information.

So what is this information about? Most importantly, each strand involves a combined program and machinery for building complementary sequences, and related polymers carry information about each others inherent structure. Thereby they also carry information about a common made of interaction with the surrounding environment, which in turn determines whether a specific sequence is replicated or not. The thermodynamic basis for natural selection (6) is thus demonstrated in an very simple manner, and it may be deduced that evolving polymers carry information, not only about their line of related structures, but also about the environment that allows them to propagate (3, 7, 13). In conclusion, evolution by means of natural selection may be understood as a thermodynamic mechanism that collects and accumulates information from thesurrounding environment (14-16) (3,8).

3. Don't stop for repair in a war zone

This physical understanding of evolution puts genetic information, and not the organism, at the center of evolutionary biology. It has direct ties to the so-called selfish gene hypothesis (7), which basically states that evolution is driven by the auto-catalytic potential of nucleic acids, regardless of any higher intentions. As the theory of evolution in general, it is provocative and controversial, but has indeed been shown to explain a number of intricate biological phenomena (17-19). We have applied this informational perspective to explore cancer development:

Cancer cells are genetically unstable (20). For every cells division they accumulate a large number of mutations, and these genetic alterations can explain the irregular morphology and growth pattem of cancer. In the beginning of 1990s it was discovered that this genetic instability is caused by mutations in genes that control and repair other genes (21). Quite logically, the instability was caused by lack of repair, and cancer development could be explained by a cascade of events in which mutations in DNA-repair genes cause mutations in numerous other genes..

However, there is one fundamental problem with this simple linear model: Cancer development implies that cancer cells proliferate and out-compete the normal cells of the body. Why then should a cell incapable of repairing its damages, grow faster then a similar, but repair proficient cell? The most common answer to this question has become that instability makes the cancer cells more adaptable. It generates potentially favorable mutations, and the unstable cells are consequently favored by natural selection (22-26).

This mutation for survival hypothesis is today the dominating dogma. It is intellectually persuasive, and appears to fit well with observations. Still, there is one agonizing problem: A random mutation is statistically much more likely to be unfavorable or lethal than it is to be favorable, simply because there are a lot more ways to destroy the genome than there are to improve it. Rephrased, each mutation may be regarded as a bet, and as for roulette the odds are always unfavorable. You may have a lucky strike, but the more bets you make, the more certain you can be to lose. There is no positive correlation between play ing big and calling out ahead, and an elevation of mutation rate will only increase the certainty for evolutionary losses. Why then should lack of repair imply an evolutionary advantage?

Several authors have tackled this problem by introducing new evolutionary principles. Evolution by sec ond order selection (27), counter selection (25) associated selection (23), and mutator hitchhiking (28) are theoretical, and to same degree mathematical models, that more or less explicitly seek to up hold the mutation for survival hypothesis. They all state that genetic instability arises, not because repair deficiency is favorable to the individual cell, but because the elevated mutation rate increases the population's overall chances for survival. In principle they therefore explain evolution of genetic instability by the controversial concept of group selection (29).

We on the other hand, have argued that cancer development and genetic instability should be understood in the above described informational perspective (30). We therefore chose the DNA repair gene as the subject, and instead of asking why instability is good for the cell, we asked why genes that repair other genes replicate poorly in cancer cells. Thereby the problem boiled down to costs and benefits of making repairs, and to illustrate this relationship we compared DNA repair genes to strategies on a race track (31):

Then to stop, or not to stop for repair, is no doubt a complex problem. Lack of maintenance increases the risk for a breakdown, whereas frequent repairs consume time and energy. Choice of strategy depends on the surrounding premises, and the intuitive conclusion is that rough environments demand frequent repairs. But that is not necessarily the case. What if the surrounding environment is quite extreme, comparable for example to a blasting war zone? Stopping for repairs would then be a deadly strategy (Fig. 3):

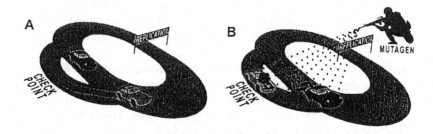

Figure 3. The 'Cell cycle grand prix', and effects of opposing repair strategies in different environments. Team I (green) always stops for repairs when a problem is indicated, while Team II (red) ignores all warning lights. Team I wins under 'ordinary' conditions (A) because it always has a faultless vehicle, while Team II accumulates errors. In the harsher environment (B) the vehicles accumulate damages more quickly than can be repaired, and Team I gets trapped in the checkpoint. Team II, on the other hand, jerks along in its faulty way with a fair chance of making the finish line. This simple assessment of repair strategies thereby provides an explanation for the paradox that mutagenic environments favor repair deficiency.

Our somewhat surprising conclusion was therefore that even though rough environments increase the demand for DNA-rep air, they would also increase the advantage of repair deficiency (31). Mutagenic environments might therefore explain the evolutionary loss of repair in cancers, and in the literature we found numerous correlations between genetic instability and the tumors' chemical environments (32, 33).

In particular, we found an intriguing relationship between the lost repair mechanism and the location of cancers in the large intestines. Whereas proximal tumors tended to loose the ability to repair small irregularities in repetitive nucleotide sequences, distal tumors were deficient in mechanisms that control the number and integrity of chromosomes. These findings were then compared to epidemiological data that relates proximal and distal cancers to environmental factors like diet and smoking, and in accordance with our assumption there was a systematic relationship between mutagenic exposure and the lost repair mechanism. In conclusion, we therefore hypothesized that the environment directs cancer development by favoring loss of specific repair genes (32, 33).

This hypothesis was later tested by explicit experiments, and its predictability was quite extraordinary (34). Cancer cells with intact DNA repair quickly succumbed to mutagenic exposure, whereas mutants with repair deficiency and genetic instability proliferated. As predicted, mutagenic environments favored loss of the repair mechanism which presumably bad evolved to prevent it, and clearly demonstrated bow an informational perspective provides new insight to the behavior of living systems.

These findings also have direct relevance for clinical practice. Chemo and radiation therapy is basically about afflicting deadly DNA damage to cancer cells. It has therefore been believed that intact repair mechanisms would prevent effective treatment (23), but our informational perspective turns this argument upside down (31). The key therapeutic trick may actually be to trigger repair in the war zone. We have recently developed a mathematical model that explains ibis relationship and a number of intriguing phenomena related to genetic instability (submitted).

4. References

1. Lloyd, S. (2000) *Nature* 406, 1047-1054.
2. Landauer, R. (1991) *Phys. Today* 44,23-29.
3. Adami, C., Of ria, C. & Collier, T. C. (2000) *Proc. Natl. Acad. Sei. USA* 97, 4463-4468.
4. Adami, C. (1998) *Introduction to artificiallife* (Springer- Verlag, New York).
5. Alberts, B., Bray, D., Lewis, l, Raff, M., Roberts, K. & Watson, J. D. (1994) *Moleculabiology of the cell* (Garland Publishing, New York).
6. Darwin, C. R. (1859) *On the origin ofspecies by means ofnatural selection* (Murray, London).
7. Dawkins, R. (1989) *The selfish gene* (Oxford University Press, Oxford).
8. Lenski, R. E., Ofria, C., Pennock, R. T. & Adami, C. (2003) *Nature* 423, 139-144.
9. Paul, N. & Joyce, G. F. (2003) *Curr. Biol.* 13, R46.

242

10. Breivik, J. (2001) *Entropy* 3,273-279.
11. Shannon, C. E. & Weaver, W. (1949) *The mathematical theory of communication* (University of IllinoPress,Urbana).
12. Yockey, H. P. (2000) *Comput. Chem.* 24, 105-123.
13. Maynard, S. J. (1999) Q. *Rev. Biol.* 74, 395-400.
14. Kauffman, S. A. (2001) *Ann. N. y Acad. Sci.* 935, 18-38.
15. Otsuka, 1 & Nozawa, Y. (1998) *J Theor. Biol.* 194, 205-221.
16. Lahav, N., Nir, S. & Elitzur, A. C. (2001) *Prog. Biophys. Mol. Biol.* 75, 75-120.
17. Goddard, M. R. & Burt, A. (1999) *Proc. Nat!. Acad. Sei. U S. A* 96,13880-13885.
18. Burt, A. (2003) *Proc. R. SaG. Lond B Biol. Sei.* 270, 921-928.
19. Dawkins, R. (1994) *Nature* 368, 690-691.
20. Lengauer, C., Kinzler, K. W. & Vogelstein, B. (1998) *Nature* 396, 643-649.
21. Kolodner, R. D. & Marsischky, G. T. (1999) *Curr. Opin. Genet. Dev.* 9, 89-96.
22. Giraud, A., Matic, 1., Tenaillon, O., Clara, A., Radman, M., Fons, M. & Taddei, F. (2001) *Seience* 291, 2606-2608.
23. Blagosklonny, M. V. (2001) *FEBS Lett.* 506, 169-172.
24. McKenzie, G. J. & Rosenberg, S. M. (2001) *Curr. Opin. Microbio!.* 4, **586-594.**
25. Janin, N. (2000) *Adv. Cancer Res.* 77, 189-221.
26. Janies, D. & DeSalle, R. (1999) *Anat. Rec.* 257, 6-14.
27. Tenaillon, O., Le Nagard, H., Godelle, B. & Taddei, F. (2000) *Proc. Nat!. Acad. Sei. USA* 97, 10465-10470.
28. Shaver, A. C., Dombrowski, P. G., Sweeney, 1. Y., Treis, T., Zappala, R. M. & Sniegows_i, P. D. (2002) *Genetics* 162, 557-566.
29. Mayr, E. (1997) *Proc. Nat!. Acad. Sei. U. S. A* 94, 2091-2094.
30. Breivik, J. (2000) On the evolution of cancer - an experimental and theoretical analysis of coevolution between genetic and none-genetic replicators in colorectal carcinogenesis. Thesis, University of Oslo.
31. Breivik, J. (2001) *Proc. Nat!. Acad. Sei. USA* 98, 5379-5381.
32. Breivik, J. & Gaudemack, G. (1999) *Adv. Cancer Res.* 76, 187-212.
33. Breivik, 1. & Gaudemack, G. (1999) *Semin. Cancer Bio!.* 9,245-254.
34. Bardelli, A., Cahill, D. P., Kinzler, K. W., Voge1stein, B. & Lengauer, C. (2001) *Proc. Nat!. Acad. Sei. USA* 5770-5775.

INDEX

A.T. Skjeltorp and A.V. Belushkin (eds.), Forces, Growth and Form in Soft Condensed Matter: At the Interface between Physics and Biology, 243-244.
© 2004 *Kluwer Academic Publishers. Printed in the Netherlands.*